AADCU PUBLICATION
WWW.AADCU.ORG
INFO@AADCU.ORG

ACKNOWLEDGEMENTS
This publication has been made possible with the help and cooperation of many individuals and institutions. Grateful acknowledgement is made to n!studio for his inspiring work and for his kind support in the preparation of book on n!studio for the AADCU Arch-Design Program.

PROJECT DIRECTOR: BRUCE Q. LAN
COORDINATOR: Simone Bove
CURATOR/EDITOR IN CHIEF: BRUCE Q. LAN
BOOK DESIGN: n!studio
EDITED AND PUBLISHED BY: Beijing Office/United Asia Art&Design Cooperation/bj-info@aadcu.org
IN COLLABORATION WITH: n!studio /www.tiscali.it; d-Lab&International Architecture Research; °C Office®

ISBN: 978-7-5609-4835-5

编辑与出版：亚洲艺术与设计协作联盟\美国\info@aadcu.org
协同合作：n!studio \意大利;
国际建筑研究与设计中心\美国;零度事务\美国

学术项目总策划：蓝青
国际总协调：罗小渠\斯坦福大学
书籍设计：n!studio

※美国亚洲艺术与设计协作联盟的建筑出版项目旨在关注国际前沿建筑学研究和实践，由于诸多因素的制约和学术环境的差异，中文阅读建议敬请读者最大程度参阅相对应之英文，并致以歉意和谢意。※

THE SENSE OF MATTER

A ADCU 2008 ARCHITECTURE DESIGN MONOGRAPH SERIES:

意大利当代建筑实践　物质感官　n!studio　Roma, Italy

生态罗马 · 弹性与极限

n!studio

critical writings

projects

essays

critical writings

The sense of matter

by Susanna Ferrini, Antonello Stella

It is strange how little we actually resemble ourselves. While this book is a collection of the most important work produced by n! studio, in all of its various configurations, between 1990 and 2006, in reality, its roots are to be traced farther back in time, to the years when we studied at the faculty of architecture in Rome. This is when we developed what, in 1990, officially became known as n! studio di architettura.
At the time abbreviations and codes were a rarity. Architects preferred to be recognised by their first and last names.
However, we were more interested in the idea of design as a collective process, where singularity, the ideas of the individual, could be 'lost' in the search for a shared meaning; we were not interested in discussions of language or formal research. Our individual and private affections for a world of forms gave way to heated discussions of the meaning of each choice made. We never spoke of 'beautiful' or 'ugly' works of architecture, but of whether architecture had or was devoid of meaning. This was part of a careful understanding of 'why' we made the decisions we did.
Each project was, and continues to be, the only difference being the number of projects involved in the same period of time, a demanding 'trial of meaning'. We have always held that the idea, interpreted as the 'meaning' of the project, is more long-lasting than its form and that it represents the only possible means to a shared and collective experience of research, such as ours, in an era of the hyper-proliferation of images such as that in which we now live. Ideas are dialectic and open; forms, on the contrary, are closed and authoritarian. The only possibility for confrontation is, as a result, the world of ideas. For us the project becomes progressively more an ethical-existential process that rotates around the evolution of an idea: unity_of thought_in the diversity_of form.
Within our research a project is capable of 'travelling' and reflecting upon the distance between our world and ourselves: "We observe the world, the world is that which we observe" to quote Merleau-Ponty. Once again design is found in the efforts to continually reduce the distance between reality and its perception.
This is what led to the choice of the name n! factorial. We wished to underline the desire to seek out the exponential multiplication of ideas and concepts in the design of architecture. We were also fascinated by the inexpressible nature of the name as a reaction to the overexposure of the architect with respect to his work. In mathematics the symbol n! factorial makes reference to concepts related to combinatorial calculations, the method of determining all possible combinations of a group of elements; in parallel, we felt that the design of architecture was not the sum of ideas, but the factorial product of them, that it was necessary to identify the current combination in order to render synchronic all of the possible configurations related to the resolution of a design theme.
The inexpressible quality of the name n! studio was intended as an allusion to an existence at the 'limit' of the design of architecture; something that was no longer the resolution of the problem of the 'form of objects', as

much as their perception and of the matter of which they are composed, observed in its mutability and evolution. This spirit, even now that n! has assumed different configurations, one of which is represented here, continues to survive in our ongoing connections with our collaborators, a close group of consultants, scholars and those responsible for cultural and museum programmes who, during this period, have contributed to an open network of research that investigates architecture and our existence in the world.

During our university years we sought references and inspiration in our recent past, in those who chose to maintain ties with Modernism, for example Franco Albini and Carlo Scarpa, so unlike the vast majority of Italian architectural research that, for many years, relegated experimentation with new materials, techniques and dimensions of architecture to a strange sort of limbo. We were interested in the perception tied to new materials, the idea of building and industrial processes created by the passage from the modern to the new forms of the contemporary. We progressively ceased to consider buildings as stable and finite forms, but rather as prefigurations captured in all of their temporariness: the momentary halting of a process. In this sense we felt that it was natural to look, in many cases beyond the Italian situation, at experiments that existed at the 'limit' between architecture and other fields, such as philosophy, art and industry, and within which the concept of temporality would allow for a consideration of architecture in all of its transformability and evolution, in its being matter in continuous transformation.

One of the themes encountered here was that of man's perception of the exterior world. We were interested in the possibility of amplifying, or even modifying it, with the objective of creating an active relationship between a work of architecture and its users. By beginning with the identification of basic and even very abstract ideas, this 'conceptual' approach to design allowed us to initiate a discourse that remains in continuous development between individuality, knowledge and a wide range of skills. The flexibility and transformability of architecture has always been at the centre of our research, based on the objective of restoring creativity to individuals and stimulating the exploration of the environment in which we live.

Other than the search for "new statutes of matter", we were also interested in the public and collective dimension of architecture. Our first occasion to confront this theme came with the design of the new Archaeological Museum in Pitigliano, which was immediately followed by the River Museum in Nazzano. A continuous relationship with the processes employed by steel fabricators and glass manufacturers afforded us the opportunity to verify the existence of a degree of continuity between research and construction: the 'ideation' of matter. This period allowed us to concretely observe the passage towards a new organisation of the office, given that for us the differentiation between research and professional practice never had any meaning, a process that generated our current configuration, established in 2002.

Sixteen years and a significant number of projects, many of which are not presented in this book, oblige us in some way to take stock of our origins, and relate the history of our work to that of Italian architecture during the same period.

Still today, and perhaps even more so, it is the meaning of architecture that interests us. This is obviously not to be understood in the general and absolute sense, but as the many partial meanings that the architect must consider in order to protect the irrefutable ethics of design research; the sense of place, his or her own understanding, interwoven with that of the others involved in the creation of a work of architecture, the sense of matter, the basic, and anything but 'inanimate' building block of architecture. There is also the sense of scale and the sense of reality, which must, however, move in time with the sense of possibility. In short, within our design process things take on form, at a certain point, without any predetermined idea or desire to move in a given direction. For this reason, as we look back over the projects presented here, we can retrace recurring themes, from the relationship with the ground plane to the continuity and fluidity of perception, through to the use of favoured materials, which assume various meanings in various projects.

However, we do not believe that it is possible to talk of 'style'. This may aid us in understanding the title of this brief premise: our projects do not resemble one another in terms of their exterior appearance, as merely apparent forms. Their similarity can only be found after a more profound reading of our work. What interests us is the quality of the perceptive stimuli and the uses that a project is capable of determining.

Our research undoubtedly features a number of recurring themes, synthesised here in broad terms as: Theme 1 – The minimum possible amount of matter, though which we seek to negate, as much as possible, the idea of matter; Theme 2 – The minimum distance between ourselves and the world, something that is achieved by increasing the potential of the tools that are capable of reducing, to the greatest degree, the distance between the observer and the object being observed; Theme 3 – The distance between reality and its perception: a reflection on the perception of objects. On the other hand, the work of architecture is something that, second only to its purely 'visible' aspects, manages in some way to 'demonstrate' its 'sensitivity'. We could say that architecture remains within the paradigm of 'apparent form' that, in other fields, such as philosophy and art is, in fact, currently being surpassed or at the very least, in a state of crisis. However, we are not looking for form. The general meaning of design wishes to be freed of the constrictions of a formal statute, in order to have an effect on the connections between objects, and not on the objects themselves. It is a process of annulment, the renouncing of a unity of configuration, in order to re-read the elements of

composition separately from one another: light, weight, and matter. From time to time it is necessary to cancel or reset the connections of consequentiality, in order to redefine new interactions between form and matter, to provoke the 'loss of meaning' that must precede any new signification. The effort is that of subverting consolidated systems for architectural elements, from the connection to the ground to the theme of openings and the skin, in order to redefine new relationships and inventions.
The relationship with matter is central to our work, with 'the ideas held by matter', to paraphrase the question raised by André Hermant in his splendid 1959 publication on architectural theory entitled Formes Utiles, to which the title of our book owes a great deal. We are talking of matter and not material: while the latter indicates a quantity of determinant characteristics, matter retains a sense of possibility and possible configurations. Musil would speak of the 'concept of the forest', in opposition to the amount of wood derived from it. Matter remains unexpressed temporality, the suspension of configuration; thus our research moves in the landscapes of abstract matter, whose surfaces tend not to contain the continuity of dimensional references, but rather references to other dimensions, from the very much smaller to the much larger...
Our design research seeks to avoid any formal drift, using the site or existing elements to closely confront the themes of space, light and matter.
Imagining matter thus means that the matter of which architecture is composed, or better yet, its perception, is that which allows the animate nature of mankind to come close to architecture. Not by chance the phenomenological approach is that which most interests us. Without wishing to create simple short-circuits with the philosophical world, all too often plundered by architects, we owe a great deal in speculative terms to the world of Husserl and Merleau-Ponty. However, these speculations run parallel to our ideas of architecture, they are part of a strong relationship that is never made explicit in direct terms. "The objects in front of us are not simple, neutral objects that we contemplate; for us each one symbolises a certain type of behaviour, recalling it, creating in us a reaction that is either favourable or unfavourable, and it is for this reason that man's tastes, his character, the attitude that he has towards the world can be read in the objects that surround him, in the colours that he prefers, in the spaces that he chooses to take a walk..." (Maurice Merleau-Ponty, Causeries, 1948)
Igor Stravinsky spoke of the creative act as a necessary and vital 'collision' with reality: there can be no creative force, and thus no thought, without the will and the ability to work in an environment constrained by rules and obstacles. During the design process we cannot ignore the passing of time; once the project is complete, it acquires its own autonomy with respect to the original design, interacting with the same reality that has generated it and transformed it over time. The process that leads to change is rendered legible, form becomes instable, while the idea, the origin of the perturbation introduced in reality will always remain perceivable. The best work is that of a line traced while walking, project-existences that cast no shadow on the ground. However, they are not inexistences, but subtle membranes that filter and envelop the existing.
This is why we do not like to speak of our work, or worse yet to describe it, something that we have left to Giovanni Corbellini. We have delegated the autobiographic description with the role of synthesising our research in the field of architecture, partly because we believe that architects give their best when they describe themselves, rather than their work, as demonstrated by so many books in this sector. Perhaps the best example is Aldo Rossi's Autobiografia Scientifica, which we mention here to demonstrate that an architect who would seem to be so distant our work can, in the world of apparent form, in the world of ideas, be so close to us.

物质感官

苏珊娜·费里尼和安东内洛·斯特拉
Susanna Ferrini, Antonello Stella

非常奇怪的是，我们很少模仿我们自己。

本书是n!studio在1990-2006年期间的最重要作品的专集，还包括各种各样的装置。在现实中，其根源还要追溯到更早的时间，直到我们在罗马建筑学院学习的时候。正是在那个时候我们提出了在1990年以n!studio di architettura的名称被世人所知的理念。

在当时，缩写和编码还非常少见。建筑师们都希望人们通过他们的名和姓来认识他们。

但是，我们对共同进行设计这一构思非常感兴趣，如果使用单一名称，在寻找某一具有共同分享意义的成分时就不能找到某一个人的构思，我们对语言或正式研究方面的讨论不感兴趣。我们对这个充满形式的世界的情感在对做出的每一个选择的意义所进行的激烈讨论面前做出了让步。我们从来不说某件建筑作品"美丽"或"丑陋"，而是说该项建筑作品是有还是没有意义。这是了解"为什么"我们做出了某项决议的过程中的一部分。

每个项目都是"对意义所进行的严格测试"，而其唯一区别就在于同一时间段内所进行项目的数目。我们一直认为，被解释为项目"意义"的构思比其形式更能经受长时间的考验，而且在我们所处的这个影像技术超速扩散的时代，它是进行共同集体研究实践的唯一可能途径，就像我们所采用的一样。构思具有辩证性且是公开的，而形式则封闭且专制的。结果，正视挑战的唯一可能的结果就是产生一个充满构思的世界。对我们而言，项目越来越多地变成了一种围绕某一构思的演变而进行的道德存在的过程：将思想统一在多样的形式之中。

一个项目能够"移动"并反映我们同我们所处的世界之间的距离。正如Merleau-Ponty所说的，"我们观察世界，而世界就是我们所观察到的世界"。我们又一次发现，设计就是持续努力缩短现实及我们对其感知之间的距离。

这就是我们选择n!阶乘这个名称的原因。我们希望能够在建筑设计过程中使构思和概念出现指数增长。我们同时也被该名称作为建筑师过度专注其从事工作的特性所深深打动。在数学中，n!阶乘为与组合计算确定某一组成分中所有可能的组合方法提供了相关的概念参考。同时，我们觉得建筑设计并不是构思的叠加，而是它们的阶乘结果。所以需要确定当前的组合，以使所有与某一设计主题的解决方案相关的可能配置同步。

n!studio这个具有无法形容特质的名称旨在隐喻建筑设计中所存在的一种"界限"，这已不再是其突变和演变过程中所出现的"物质的形式"这一问题的解决方案了，对它们的感知以及它们的组成物质也同样如此。即使现在n!已经具有了不同的配置，这种精神在我们同我们的合作者继续联系的过程中仍然存在，我们的合作者们包括顾问、学者和负责文化和博物馆规划的人。在这一时期，他们在为我们创建一个开放型建筑研究网络以及帮助我们在世界范围内发展方面作出了突出的贡献。

在我们的大学时光中，我们在刚刚经历的事情中，借助如Franco Albini和Carlo Scarpa等与现代主义保持着联系的人中寻找参考和灵感，这与很多年来都放弃进行试验而只追求新材料、新技术和新规格的大部分意大利建筑研究机构均不同。我们对与新材料相关的感知、建筑的构思和从现代到当代的新形式的转变过程中所产生的工业过程感兴趣。我们渐渐不再认为建筑是稳定、有限的形式，而是将它们看作是具有各种临时性属性的原型：即某一过程中临时的停顿。在此意义上，我们看待那些位于建筑以及如哲学、艺术和工业等其他领域"界限"上的试验就显得很自然了，而很多情况下这些案例都是在意大利之外的案例。在此其暂时性的概念使得我们能考虑建筑的所有转变性和演变性，使其成为一种持续变化的事物。我们此处所遇到的一个主题就是人对外界世界的认知。我们对将其放大甚至是更改非常感兴趣，希望能够在建筑作品及其使用者之间建立一种积极的联系。从确定基本的甚至是非常抽象的构思开始，这种"概念性"设计方法使我们能够在个性、知识和多种技巧方面保持持续发展。建筑的适应性和可转变性一直是我们研究的核心，并以恢复个体创造性以及促进对我们所居住环境的开发这一目的为基础。

除了寻找"新的物质规则"之外，我们还对建筑的公共和集体特性感兴趣。在我们第一次面对这一主题时，我们设计了Pitigliano的新考古博物馆，而它马上便被Nazzano的河流博物馆所效仿。钢铁制造商和玻璃制造商与其工艺流程所保持的持续关系为我们验证和研究建筑之间的连续性提供了机会：这就是事物的"构思过程"。该阶段使我们能够实际观察工作室重新组织的过程，对于我们而言将研究和专业作业部门分开从来就没有什么意义，正是这个过程造就了我们当前所采用的配置形式，也就是2002年所建立的形式。

16年的经历以及我们所完成的众多项目，在某种程度上促使我们来回顾我们的起初，将我们的工作史与同时期的意大利建筑史结合起来。

直至今天，我们所感兴趣的仍然是建筑的意义，而且这种程度甚至更深了一些。显而易见，从一般意义上和绝对意义上讲这是不可理解的，但是与建筑师为了维护其设计研究工作中不可辩驳的道德规范而必须要考虑的许多东西一样，对空间的感觉，对事物的感觉，还有对规模的感觉和对现实的感觉，这些必须与对可能性的感觉同步。简而言之，在我们的实际过程中，事物会在某一特定点，在没有任何预先决定的构思或向某一特定方向移动的愿望的情况下，悄然成形。因此，当我们回顾此处所述项目时，我们能够重新追溯重复出现的主题，从与地平面的关系到感知的连续性和流动性，再到使用所喜欢的材料，而它们在不同的项目中具有各种不同的意义。但是，我们并不认为我们能够探讨“风格”问题。这可能能够帮助我们理解这篇前言的标题：我们的项目在外观，即仅仅在明显形式方面，相互之间没有相像之处。只有更加深入地阅读了我们的作品才能够找到它们之间的相似之处。我们所感兴趣的是感知上的刺激以及某一项目所能呈现的用途。毫无疑问，我们的研究特别描述了一些重复出现的主题，在此处合称为：主题1——最少量的物质，虽然我们想尽量排除对物质的考虑；主题2——我们本身同世界之间最短的距离，这一点是通过增大能够最大程度减少观察者和被观察物品之间的距离的工具的能力来实现的；主题3——现实及其感知之间的距离，即对物品感知的反映。另一方面，建筑作品是一种在一定程度上努力“展示”其“敏感性”的东西，仅次于其纯“可见”方面。我们要说，建筑仍然停留在“表观形式”的规范之内，而这种形式相比哲学和艺术等其他领域，现在已经过时，或至少是处于危机之中。但是，我们所追求的不是形式。设计的一般意义希望能够摆脱形式规则的束缚，以对物品之间的联系，而不是物品本身产生影响。这是一个废除的过程，是对某一配置个体的放弃，以相互分离的形式重新诠释某一复合体的各种成分。就像光、重量和事物一样。我们需要不时地取消或重新设置贯性之间的联系，以重新定义形式和物质之间新型的相互关系，激发在具有任何新意义之前必须完成的“丧失意义”过程。其努力方向在于为建筑元素推翻固定的系统，从与地面的联接到开口和皮肤的主题，以界定新的关系和发明创造。

与事物的关系对我们的工作而言非常重要，通过“事物所具有的构思”，来回答Andre Hermant在1959年就建筑理论所发表的题为《Formes Utiles》的杰出论文中所提出的问题，而我们这本书的书名在很大程度上借鉴了这篇论文。我们所谈论的是物质，而不是材料，后者能够显示大量的决定性特性，物质则能够保留一种可能性和可能配置的感觉。Musil则会说这就是“森林的概念”和从其中得到的大量木材之间的区别。事物可保留并未说明的暂时性，即悬而未决的配置。因此，我们的研究便是驰骋在抽象事物的场景之中，其表面一般并没有空间参考的连续性，而是有对其他规格的参考，从小得多直到大得多……

我们的设计研究努力避免任何形式上的偏差，通过将场地或已有的元素更为密切地对比空间、光线和物质等主题。

因此，想象物质就指的是构成建筑物的物质，或更恰当地说，其感知使人类的充满活力的特性同建筑物结合在了一起。绝非偶然地，我们最感兴趣的就是现象学的方法。我们并不希望同哲学世界建立简单的联系，这一领域太频繁被建筑师们争夺。在推理方面我们在很大程度上应感谢Husserl和Merleau-Ponty的世界。但是，这些推理同我们的建筑构思并行，它们是一种从来没有被直接阐明过的关系中的一部分。“在我们面前的物品并不简单，它们是我们所期待的中性物品；对我们而言它们每一个均代表一种类型的行为，对其进行回顾，能够对我们形成一种有益或有害的反应，正因为如此，我们能够通过一个人周围的物品，其所喜欢的颜色、喜欢散步的地点等来判断其品位、性格，及其对待世界的态度……”（Maurice Merleau-Ponty，文艺随笔，1948）

Igor Stravinsky说创造性行为是同现实之间必要性和致命性的“碰撞”：世间可以没有创造力，因而也就没有思想，在受规则和障碍约束的环境中没有工作的意愿和能力。在设计过程中，我们不能忽略时间的推移，项目完成之后，它便获得了其原始设计的自主权，同创造出它并随时间将其改变的同一现实相互作用。导致变化的流程变得非常明显，形式变得不稳定，而构思，向现实中引入动乱因素的源泉，将总是可以理解的。最好的作品就是散步时所追溯的一条线，项目的存在并没有在地面上留下任何阴影。但是，它们也并非不存在，只有通过过滤和剥开存在于事物表面的精细薄膜。

这就是为什么我们不喜欢谈论我们的作品的原因，或更不好听地说，那时我们留给Giovanni Corbellini要解决的事。我们将自传描述部分内容用于对我们在建筑领域的研究进行综合描述，部分原因是因为我们相信建筑师们对自己而不是他们的工作，进行介绍时能够将其最好的一面展示给大家，这如本领域许多著作已经证实的那样。其最好的范例也许就是 Aldo Rossi 的 *Autobiografia Scientifica*，我们在此引用该著作是为了证明，建筑师似乎离我们那么遥远，但其作品，在这个表观形式的世界，这个构思的世界，可以距离我们很近。

A Small En!cyclopedia

by Giovanni Corbellini

It is difficult to speak about the **abstraction** of architectural design. At least it is much less clear than in the world of painting, where demarcation through figurative approaches is conceptually described and widely used. Often the use of this term is limited to capturing geometric similarities with neo-plastic, suprematist or constructivist explorations, neglecting the reasons and processes that have led to a particular architectural result. It is a process that the work and designs of n!studio are particularly effective at demonstrating. The abstraction that we can observe in their work is not presented as a preconception, an a-priori form, but as the result of a series of successive steps, within which the fundamental elements of design are progressively freed of their more evident and constructive, representative and symbolic necessities. The design strategies adopted by Susanna Ferrini and Antonello Stella define the emergence of a sort of desire for mimesis, which takes advantage of the particular aspects of a condition or context, turning them into a source of inspiration for sublime and rarefied solutions. The boxes full of archaeological remains that become articulated sheets of cor-ten steel in the storage building at the Villa dei Quintilli, or the plastic tents used to protect lemon groves that are transfigured into the roofs of the Tourist Centre in the Valle dei Mulini in Amalfi are like instruments of environmental adaptation, whose successive 'homeopathic' dilution renders the original morphologies almost unrecognisable, if not at the level of the sublime. This process of abstraction also affects the basic elements of construction. Walls, roofs and windows, together with the other 'archetypal' elements of architecture, dissolve and expose the questions that lay at the base of their very existence; supporting, protecting, insulating, opening, connecting, etc. The solutions adopted introduce a new repertoire of forms made of surfaces that are folded, cut, overlapped and oscillating. This need of overcoming the **banality** of 'correct' and consolidated typologies come to the fore with great force, above all when it is most difficult to avoid it, where the 'normality' of uses and/or pressures of a morphological adaptation to existing environments is highest. The cultural centre in Ciampino concentrates a vast quantity of solutions within a small building, making it exemplary of the attitude adopted by n!studio in these cases: while the roof responds to the necessary requirements of slope, it does so by overturning the standard profile of the roof; it is virtually isolated from the volume below by a strip of glazing and an attentive focus on the geometric variations of the steel supports, in turn separated from the concrete piers that divide the spaces inside. The traditional profile of the roof is, vice versa, reinterpreted in the Centre des Eaux Douces-Reunion, where it is transformed into a continuous and variable element that, embedded at a certain point in the earth, erases its typological origin. The use of folding ribbons to resolve the exterior envelope can be found in many of the office's projects, including the poly-functional centre in Morena, Amalfi, the Porta Susa Rail Station in Turin, in Marseille amongst others. The reference to elements that are found in the local **context**, whether artificial, as described above, or natural, on the one hand resolve the necessity of satisfying the need for simple images that are symbolically effective, directly understandable by clients, administrators and users and, on the other hand, expressing a sincere desire to root the project. The design process that produces these themes also leads to a progressive critical **detachment** that is made manifest in the contrast between the nature of the subjects and their geometric simplification, their scalar extension and, above all, the hyper-artificial substance of their construction: the stainless steel 'cane thicket' in the museum in Nazzano is rendered gigantic in the concrete columns wrapped in the methacrylate panels of the pumping station in Valle Giralda; the iridescent luminosity of water is pursued on many occasions in the variable reflections of glass; the light that filters through the trees inspires the etched 'branches' that cover the fibreglass skin of the high school in Rome. This attention towards the morphologies of nature is accompanied by a more intense adhesion to its dynamic processes. The buildings of n!studio exist in time, to the point of assuming an **elasticity** that becomes one of their principal design parameters. Configurations and 'variable geometries', used both for exhibitions and residential experiments (the addition to the GNAM - National Gallery of Modern Art in Rome, the Tourist Centre in Amalfi, the Taquin House, etc.) all propose sophisticated technological solutions, together with a vital adaptability to changing situations. The **factorial** number, the raison d'etre of the office, thus becomes immediately understandable: other than synthetically describing something

grand, it is also a mathematical instrument for representing the uncertain and proliferative phenomena of nature, and as such it is used in quantum mechanics and to calculate probabilities. The paradoxical technological reinterpretation of natural events pursued by n!studio is as far from **hi-tech** determinism as it is attentive to processes of building **industrialisation** and the effects that they can induce. This generates a design approach that presents as little as possible of the real substance of construction. The continuous surface of the wall is used to hide the beams that support the cantilevered stair treads in the museum in Pitigliano, while the complex mechanisms that provide access to the display cases in Nazzano are concealed behind the weightless immateriality of etched glass. Themes and **locations**, a sense of spatial design and continuity, are privileged – through an authentically Italian Modernism – over the more tectonic and self-referential characteristics of construction. It is once again the conditions of location, and not a desire for animation, that is purely gestural, that modifies the structures of composition, introducing important deviations from basic geometric schemes. The use of parallel strips to create depth, change dimensions and shift from one to the other (what could be called n!studio's 'signature' can be found in the Borghetto Flaminio project through to the new high school in Rome, including the GNAM, Amalfi, Marseille, the piezometric tower, etc.) develops in an interesting manner in the project for the poly-functional centre in Morena, where the necessity of vertical circulation, tied to the irregular forms of the site, forces the strips to bend, penetrate one another and deform in the various directions of the building spaces. Here – as in the Olympic Village, the Fiumarella Park in Catanzaro and the Enel Park in Civitavecchia – the operations have a significant impact on the ground plane, which becomes, for all intents and purposes, the physical and conceptual **matter** of the ideation. It is an approach that finds its most complete resolution in the recent competition-winning proposal for the new archaeological museum in Breteuil, where the design of the ground plane constitutes the fundamental instrument for the insertion of the project within the landscape, introducing a slight wrinkle that underlines the delicate conformation, without interrupting its continuity. Even in this case, as with the other construction solutions, the desire **not to show** reveals a precise and unexpected strategic approach that overlaps a second, evanescent and elusive level of reading the **object-like** substance that is so present in the architectural considerations made by n!studio. The tension that is created between spaces and movements, the refined ambiguity of the existing elements, which are made to disappear and the ability to float heavy objects speak of a sophisticated ability to interact with the expectations of observers, attracting their interested and surprised glance.

Perception is thus a specific part of the research carried out by Susanna Ferrini and Antonello Stella who, evidently, do not limit themselves to pursuing the purely retinal beauty of materials, of surface patterns, volumetric articulations and their composition in more or less balanced groups. Thought and matter, vision and empathy contribute, for example, to supporting the peculiar **quality** of the archaeological museum in San Vincenzo al Volturno. This is a singular work in the development of n!studio, partially due to the choice to wrap the common linear arrangement of elements within a spiral that is unusually decentred and partially due to the ability to concentrate a series of viewing instruments that create complex **relationships** between architectural objects and the landscape, between the objects on display and the territory to which they belong. Other than the attention with which the most interesting elements are framed inside (the arms of the spiral become optical machines that confirm the centrifugal structure, leading the eye towards the exterior) there also emerges an implicit **stratification** of spaces and materials, of their finishes and their use in terms of light. Various layers of **transparency** organise the interior articulation of the more public areas and those used solely for exhibition space, simultaneously obtaining a spatial and visual coherence that is articulated in depth. In the same way, the attenuation of the coordination between the layers of the building envelope, as if they were independent and able to slide, introduces complex overlappings of transparent, opaque and translucent elements and the external skin in burnished, perforated brass, with varying degrees of transparency. The result is a building that is visually unstable, sensible to the passing of time, changes in light and the movement of people. A conception that expresses the intention to create a continuity with the conditions of the site, as in the projects for the San Luca crocera in Brescia, the museum in Salzburg or the new rail station in Turin, which seek to create a vaster urban **unity**. The idea of rarefaction that runs through many of n!studio's design strategies (from the decomposition of building and typological archetypes to the concealing of technique, ending with the dissolving of urban and landscape continuity) finds its definitive resolution in the transparency of glass. However, the unity that is revealed is always presented in a complex manner, graduated by a succession of horizontal planes within which **vibrations** of light assume a paradoxically material consistency. Reduction is once again proposed as the new instrument of process, moving backwards through the factorial progression, without ever arriving at the negation of absence. On the other hand, the factorial of **zero** is, by definition, equal to one.

n!小百科全书

Giovanni Corbellini

谈论建筑设计的“抽象”概念非常困难。这至少不像油画世界那样明确，油画世界中采用概念性描述并广泛使用了图形方法来进行划分。通常这一术语都是仅限于描述新造型、至上主义或与建构主义研究中的几何学的相似之处，并不考虑导致形成某一特殊建筑结果的原因和过程。n!studio的作品和设计特别能够体现这种过程。我们从他们的作品中所看到的抽象概念并不是表现为一种事先构想，一种先验形式，而是一系列连续步骤的结果，通过这些步骤设计主要成分的建设性、代表性、象征性和必需性被逐步体现了出来。Susanna Ferrini和Antonello Stella所采用的设计战略阐释了所涌现的一类对模拟的渴望，它利用某一条件或背景的具体情形，将其转化为一种卓越而稀少解决方案的灵感源。

在Villa dei Quintilli的储藏建筑中，满是考古遗迹的盒子被转变为相互联结的低合金高强钢板。在Amalfi 的Valle dei Mulini旅游中心用于保护柠檬林的塑料帐篷被转化为了旅游中心的屋顶，它们即使不能算是卓越的作品，也可以算是进行环境保护调整的仪器。它们随后形成的“顺势疗法”稀释作用使得我们几乎无法辨认出其原始形态，墙壁、屋顶和窗户同建筑物的其他“原型”混合到一起，解除并暴露了其根本存在性问题，提供支持、保护、隔离、打开、连接等功能。所采用的解决方案引入了由折叠、切割、交迭和摆动的表面构成的全套新型形式。需要克服“正确”这一概念的平庸之处以及统一类型的需求，尤其是当最难避免时，即使用的“正常性”和/或需要根据现有环境对形态进行调整的压力最高时。

Ciampino的文化中心在一座小型建筑中集中采用了大量的解决方案，它代表了n!studio针对这些情况典型的处理方式：对应于屋顶必要的坡度要求，设计摒弃了屋顶的标准外形，通过金属支撑结构上几何形状不断变化的玻璃装配带将其同下面的建筑体实质性地隔离开来，依次将分割内部空间的混凝土墩相隔离。而反过来，Centre des Eaux Douces-Reunion的传统型的屋顶结构则被重新进行了诠释，转化成了一种连续且不断变化的成分，嵌入到地上的某一特定点，完全消解了其原来的形状。这种采用折叠带状结构来解决外部包封问题的方法可在该工作室的许多项目中看到，包括Amalfi的Morena的多功能中心、Marseille的Turin的Porta Susa火车站等。

对于本地环境中特有元素的引用，无论是人工形式的还是天然形式的，一方面满足了对要求具有的象征意义，客户、管理人员和使用者可以直接理解的简单形象的需求，另一方面表达了牢固建立该项目的诚挚愿望。在生成这些主题的设计过程的同时也导致一种渐进型的关键分离，将物品的天然特性同其简单的几何学特性、其数量扩增同最重要的建筑超人工成分相对比，可更明显地发现这种分离：Nazzano博物馆的不锈钢藤茎密集屏障在Valle Giralda泵站包有异丁烯酸盐板的混凝土柱的包裹中而显得非常巨大；在许多情况下，通过不断变化的玻璃反射来体现水的色彩斑斓的光影；在罗马的高级中学，透过树木过滤的光线照射在其玻璃纤维表层上，使人们想起了被蚀刻的“枝条”。在注重自然形态的同时，更加注重其动态变化的进程。n!studio所建造的建筑与时间同在，他们甚至将弹性当作了他们的一项主要设计参数。他们所采用的配置和“可变形几何空间”，既可用于展会又可用于住宅试验（罗马GNAM-国家现代美术馆扩建项目、Amalfi的旅游中心、Taquin House等），都能够提出非常尖端的技术解决方案，同时还能够对不断变化的条件保持极高的适应性。因此，我们就可以更好地理解其阶乘数，其办公室的raison d'etre除了用于综合地描述非常大的物体之外，它还是一种解释自然的不确定性和增生现象的数学计算工具，用于量子力学概率计算。

n!studio所追求的对自然事件的反论技术再阐述远不只是高技术决定论，就像它注重建立工业化的过程及其可能带来的影响一样。这就生成了一种尽可能少地呈现建筑的真实成分的设计方法。在Pitigliano博物馆，其墙壁的连续表面可作为隐藏支撑悬壁式梯级踏板的横梁，而作为Nazzano展示柜入口的复合构造则被隐藏在了无重量非物质的蚀刻玻璃后面。借鉴一种真正的意大利现代主义观点，相对于建筑工作更加倾向于建筑性和自我参考性的特点，特别彰显出了主题和位置，一种空间设计和连续性的感觉。这是位置的条件，而不是对活力的期望，属于纯姿态的问题，可以更改布局中的结构，在基本几何学方案中引入重要的偏移成分。这种通过平行的带状结构来增加深度、改变大小，以及从一个结构转移到另一个（这可以被称为是n!studio的“签名”，可以在从Borghetto Flaminio项目到罗马新高中等多个项目中看到，包括GNAM、Amalfi、Marseille、piezometric塔等）的方法在Morena的多功能中心项目中以一种有趣的模式得到深入运用。在项目中采用的一种竖直循环，由于受到该场地不规则形状的约束，而迫使这些条状结构发生弯曲，相互渗透，在其建筑空间中发生各种方向的变形。就像在奥林匹克村、Catanzaro 的Fiumarella公园和Civitavecchia的Enel公园一样，其所采取的操作对地平面有很大的影响，而它则在所有用途和目的方面成为了构思过程的物理和概念“素材”。在Breteuil新考古博物馆这项最近的获奖提案中，这种方法得到了最完美的解决，在其中地平面的设计成为了将此项目插入到其地貌风景中的重要工具，引入了一种着重突出其精巧构造的轻微波纹结构，而同时也未影响其连续性。

即使在这一案例中，同其他建筑解决方案一样，对“不显示”的期望展现了一种准确而出乎意料的战略性方法，而这种方法又同第二种、渐渐消失且难以捉摸的读物相交迭，即在n!studio的建筑设想中如此明显的“物品样”成分。在空间和移动之间营造出的压力，现有成分中经过重新定义使其消失在模棱两可之处，以及使重物体漂浮的能力，如同观察者的预期互动的深奥而微妙的能力，吸引着他们惊奇而又充满兴趣的目光。因此，感知就成为了Susanna Ferrini和Antonello Stella所进行研究中的一个特殊部分，显然，他们并不局限于追求物体、表面造型的纯视觉美感，他们还追求建筑连接以及它们或多或少的均衡组合。这是n!studio发展过程中的一项独特工作，一方面因为选择了将普通线形的成分包在异常偏心形式的螺旋结构中，另一方面是因为其将一系列的检查工具集中在建筑物品和地貌风景之间，在展示物品和其所属的区域之间建立一种复杂的“关系”的能力。除了在将最有趣的成分框在框架之内（螺旋结构的臂成为了确认其离心结构的光学机器，将眼球吸引到外面）方面注意之外，还出现了一种含蓄的空间和物品层次，包括其抛光度及其光线使用情况。各种层次的透明度将公开程度更高的区域以及纯用于展览的区域的内部连接组织起来，同时获得了深层次连接的空间和视觉一致性。按照同样的方法，建筑整体层次之间协调度衰减，就像它们均独立存在且能够滑动一样，引入了各种透明、不透明以及半透明成分之间的复杂交迭情况，且其外部表层采用抛光、带孔的铜质结构，具有各种各样的透明度。这样所得到的建筑就是一种视觉上不稳定、可感知时间的推移、能够随着光线和人流变化而变化的建筑。这是一种表示要营造一种与现场条件具有连续性的意图的概念，如Brescia的San Luca crocera项目、Salzburg 博物馆项目或Turin的新火车站项目，它追求营造一种更广度的城市“联合”。

这种贯穿很多n!studio设计战略的构思（从分解建筑和类型原型到技巧的隐藏，最后到解决城市和地貌风景之间的连续性），都通过玻璃的透明性得到了确切的解决方案。但是，其所揭示的联合效果总是以一种复杂的形式呈现，被一系列的水平平面分割定级，而光线的“震动”在其中形成了一种似是而非的物质连贯性。而删减又一次被作为此流程中的一种新工具提了出来，在其阶乘进程向后返，从来没有达到对缺失的否定。另一方面，因子是零时被定义为等于一。

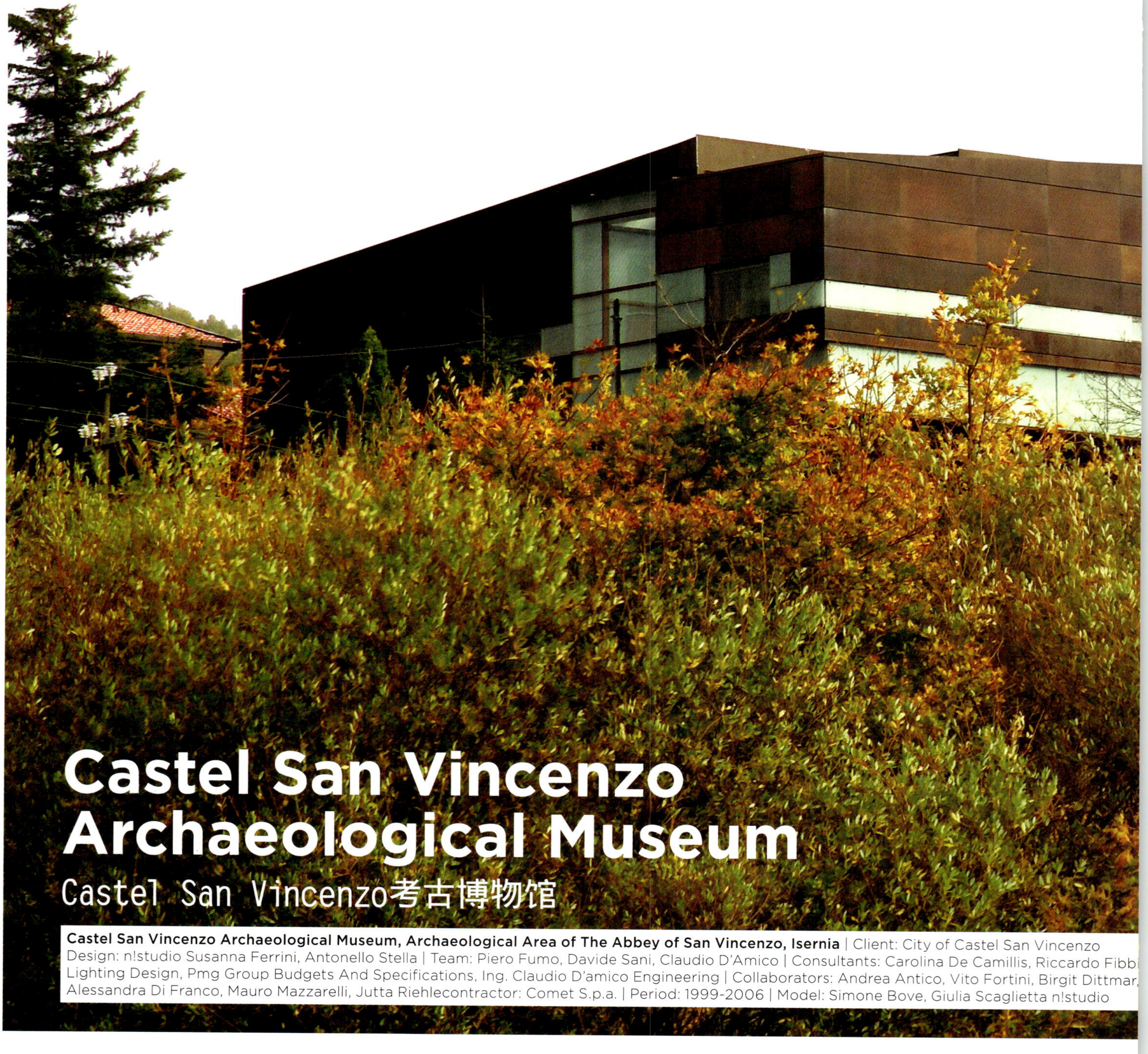

Castel San Vincenzo Archaeological Museum

Castel San Vincenzo考古博物馆

Castel San Vincenzo Archaeological Museum, Archaeological Area of The Abbey of San Vincenzo, Isernia | Client: City of Castel San Vincenzo Design: n!studio Susanna Ferrini, Antonello Stella | Team: Piero Fumo, Davide Sani, Claudio D'Amico | Consultants: Carolina De Camillis, Riccardo Fibbi Lighting Design, Pmg Group Budgets And Specifications, Ing. Claudio D'amico Engineering | Collaborators: Andrea Antico, Vito Fortini, Birgit Dittmar, Alessandra Di Franco, Mauro Mazzarelli, Jutta Riehlecontractor: Comet S.p.a. | Period: 1999-2006 | Model: Simone Bove, Giulia Scaglietta n!studio

sense of matter

The museum is dedicated to the storage and display of archaeological remains from the abbey of san vincenzo al volturno, near the head of the volturno river in the region of molise. It is one of the most important medieval archaeological excavations in europe, witness to the discovery of important and numerous remains that have helped write the history of the late medieval period. The building is a unicum of content and container, where the structure and the cladding coincide with the design of the display spaces: the display cases become tunnels that the visitor can enter, discovering different levels of perception and awareness. The architectural idea is the result of a synchronisation with the idea of presentation, such that all display systems become 'visible' on the exterior of the building. The building is a helical volume, suspended above the level of excavation, with a continuous façade in burnished, micro-perforated copper above sheets of acid-etched glass that, illuminated at night, contribute to image of the suspension of the upper volume. The lower part of the building, partially 'embedded' in the section of the earth, contains the restoration workshops, the storage areas, the temporary display spaces and a seminar and meeting room, while the second level, offering access to the municipal building, contains the permanent exhibitions related to the archaeological site.

The building is characterised by two types of display cases: the first is composed of a system of tunnels running along the spiral of the main display itinerary; visitors walk inside the display cases, in continual visual contact with the main circulation route, accessing more intimate spaces that are ideal for the further investigation of various topics and an increased level of involvement. The second system involves the design of the interior façade, the 'skin' of enclosure, composed of an alternation of display cases and transparent strips that frame the panorama of the surrounding landscape.

此博物馆专门用于保存和展示位于Molise区Volturno河源头附近的San Vincenzo al Volturno修道院的考古遗迹。它是欧洲最重要的中世纪考古挖掘项目之一，是中世纪晚期众多重要遗迹发现历程的见证。该建筑是一种容器和内容的集成体，其结构和覆层与展示区的设计完全匹配：展示柜成为了游客可以进入的隧道，在其中他们可以获得不同层次的感知和认知。这是建筑概念与展示概念同步的结果，这样所有的展示系统都可以在建筑外部变得“可见”。该建筑采用的是一种螺旋状的结构，悬在出土文物层上方，其连续的正面外表面由经过抛光、上面开有微孔的铜板覆盖在酸蚀玻璃板上构成，夜晚在灯饰的辉映下，增添了其上部结构的悬浮感。该建筑下部的一部分“埋入”地下，包括修复车间、存储区、临时展示区以及会议室，而第二层则可以连接市政建筑部分，为与考古区相关的永久性展示区。

此建筑的特色体现在两种展示柜上：第一种由沿主展示路线螺旋式建筑分布的隧道系统组成，游客可以步入展示柜中，持续享受主循环路线的视觉盛宴，进入最适宜对各种主题进行进一步观察研究的更私密空间，增加了其参与度；第二种系统涉及到了其内部表面的设计，即其外壳的“表层”，它由展示柜和透明带交替构成，可以将周围风景尽收眼底。

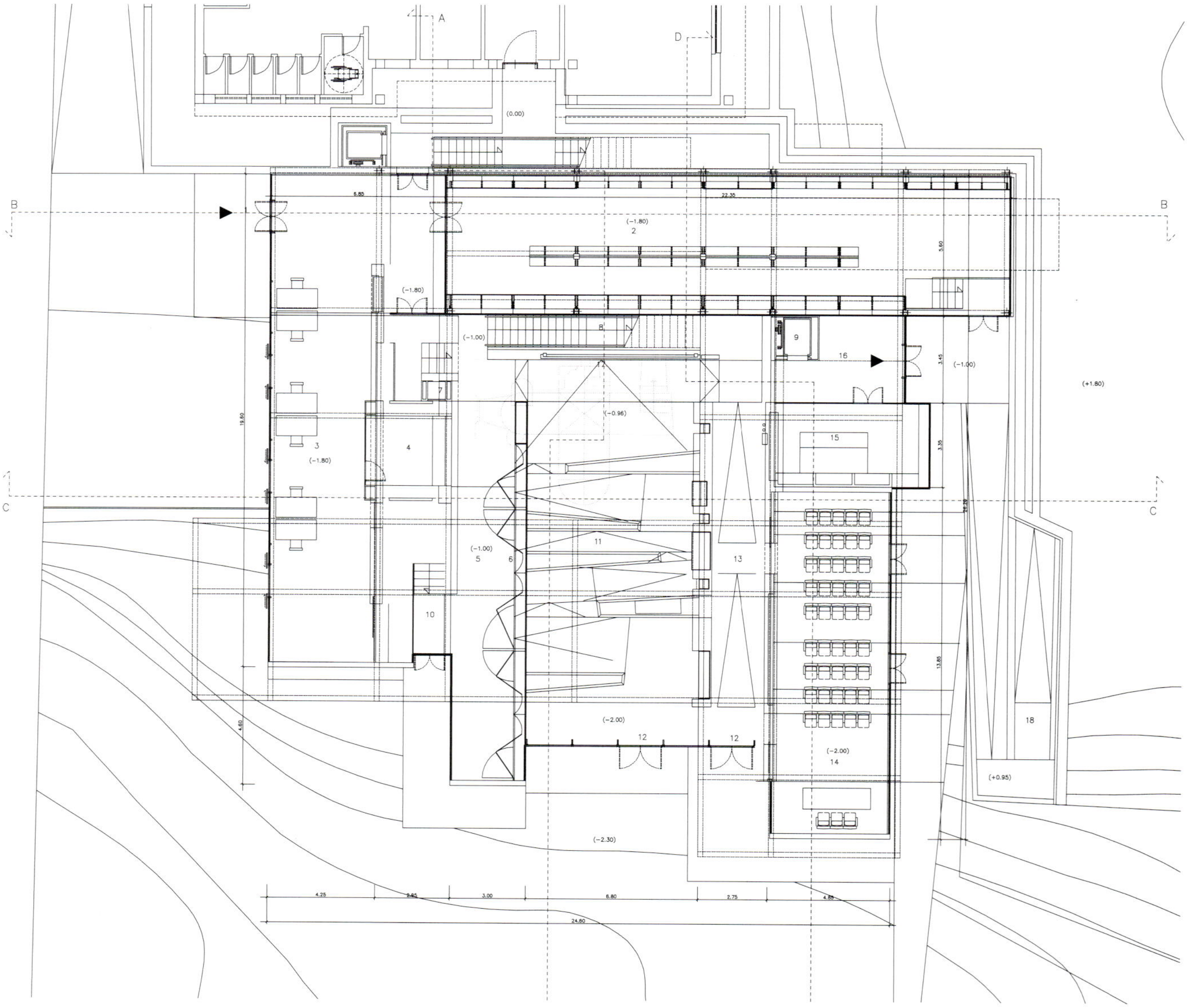

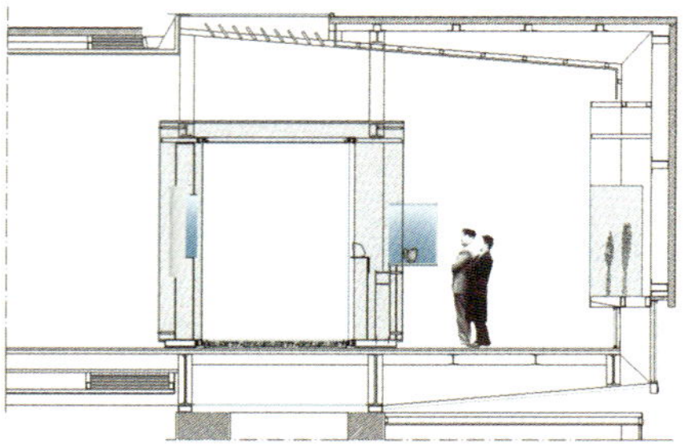
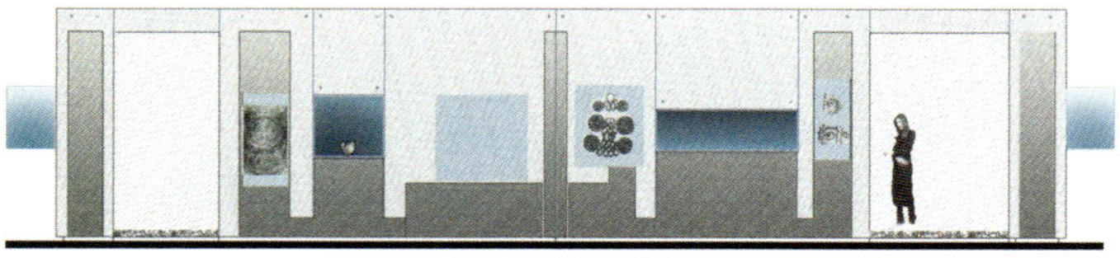
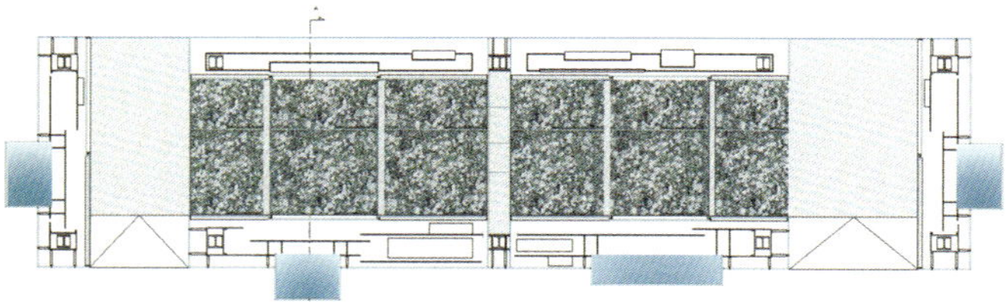
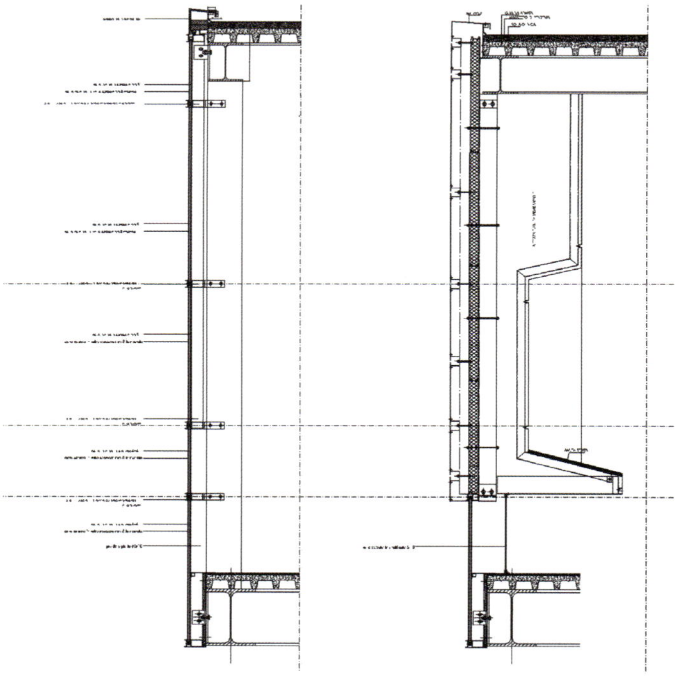
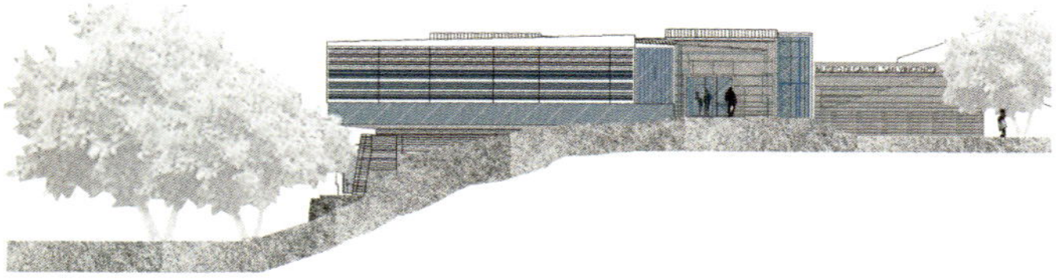

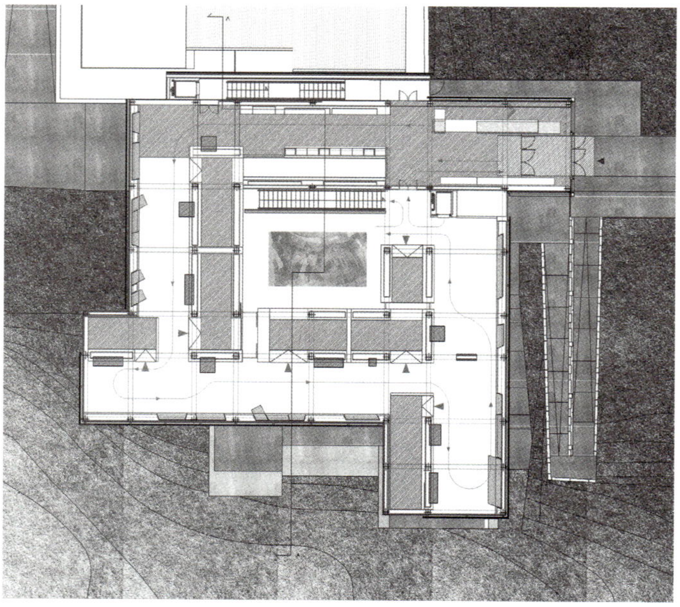
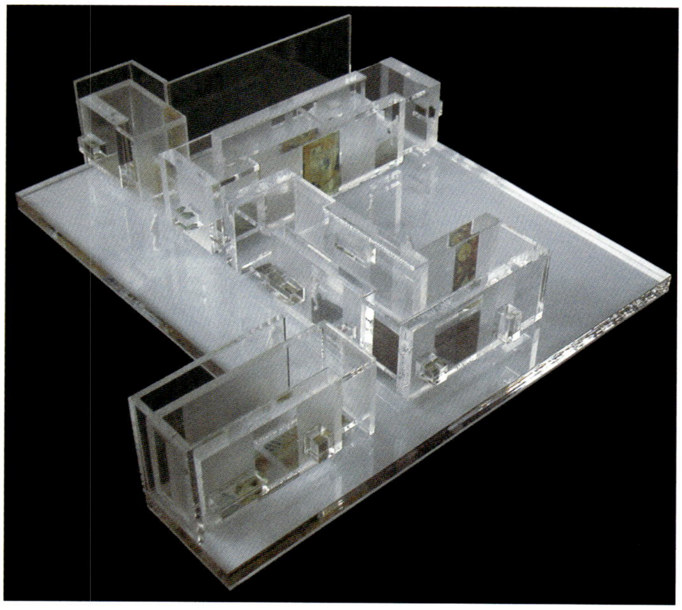

Mind the Gap

毗邻建筑

New Educational Complex and Public Park, Rome | Design: n!studio Susanna Ferrini, Antonello Stella | Team: Simone Bove, Stefano Battaglia, Maria Angela Corsi, Sara Ferazzoli, Paolo Rossi, Giulia Scaglietta | Consultants: Carolina De Camillis, Riccardo Fibbi, Gaetano Fasano, Prof. Dario De Luca | Engineering: Pr.as. Tecnica Edilizia S.r.l. | Client: Comune di Roma | Period: 2006

sense of matter

7A
7B
7C
7D
4A
1A
2A
3A
11
12

The compression of the project site in the narrow and long space between the new residential complex and ecological corridor of the Cecchignola Ravine suggested the design of an urban system capable of functioning simultaneously as a filter and element of unification between the proposed new interventions. A unique, mineral sign that ties together the entire building façade being constructed, guaranteeing a continuous connection with the new park.

A plate, slightly raised above the road level, is intersected by a series of transversal elements that, from the built fabric, insert themselves within the park as part of a slow and progressive fragmentation towards the countryside.

In this case the line of separation between the natural and the artificial is rendered more evident, materialising in a 5-meter level change that divides the competition area in two. This physical barrier between the countryside and the city hinders the discovery and investigation of nature. The result is a critical edge situation, the limit between two contrasting realties, which continually seek new possibilities for dialogue. These two premises generate two new scholastic buildings, the ideal propagations of the city within the agro romano and extension of urban activities into the natural world. They are virtual bridges between the progressive expansion of inhabited centres and the reduction of the countryside. The artificial element extends into the countryside as part of a continuous process of folding that is penetrated by natural elements. The exterior space weaves its way into the schools that, in turn, extend out towards their surroundings. The buildings are open towards their context as part of a continuous dialogue and tension between inside and outside. By extending into the park, the buildings become one with nature, losing the closed geometry of the street façade, and progressively opening towards their surroundings, while the continuous insertion of natural elements slowly fragments their volumes.

The transformations of the buildings, their opening and closing towards the exterior, is mirrored in the articulation of the interior spaces. In correspondence with the spaces for free and regulated activities inside the classrooms, the building is closed up to a height of 1.2 metres, creating a large glazed horizontal cut that allows for a permanent

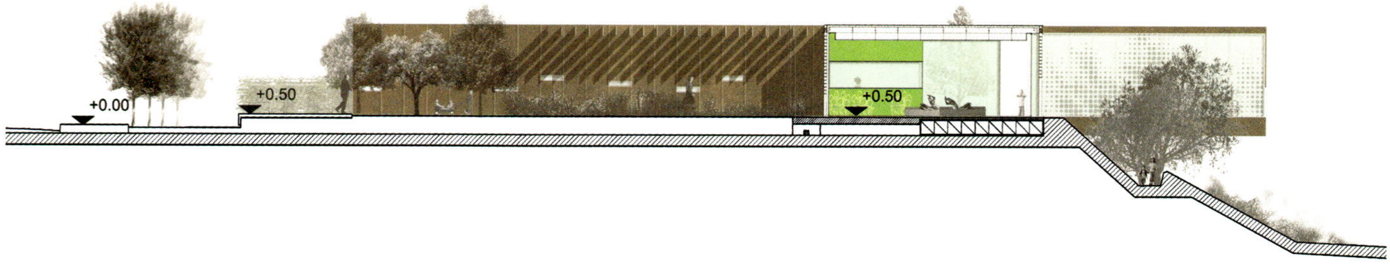

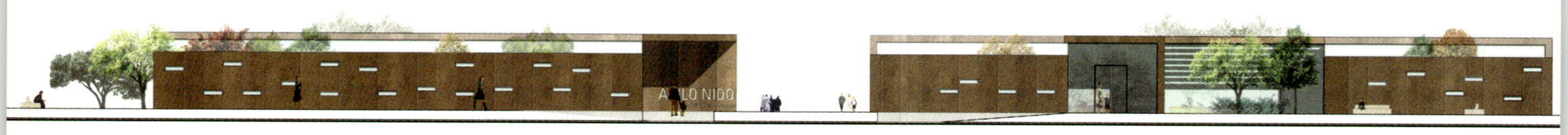

visual relationship with the exterior. The spaces of the workshops are characterised, on the contrary, by a closed wall on one side, with the exception of a few small horizontal openings, and a full-height glazed surface that projects the interior space towards the Roman campagna, defining a privileged point for the observation of this natural setting. The public park is designed together with the plate structure that runs along the entire edge of the project. The plate is characterised by the alternation of rest areas and insertions of natural plantings that offer shadow. These are the starting points of the transversals axes that cross the park and project residents into the natural setting of the agro romano. As these paths move into the park, they progressively lose their consistency, de-materialising completely to make way for the simpler signs of a well-worn earthen path. A second network of paths, made using natural materials, runs through the park and accompanies this system of connections. Its elements are differentiated according to various activities: bicycle paths, pedestrian paths, athletic paths and educational walks.

此项目的现场位于Cecchignola Ravine新住宅复合中心和生态走廊之间的狭长地带，我们提出了一种能够同时行使过滤器功能和将新增项目联合在一起的城市系统设计。一种将整个正在建造的建筑外表面结构连接在一起的独特设计保证其能够同新公园持续连接在一起。一个比路面稍高一些的板式结构同一系列的横向结构相互交错，这些横向结构从建筑构造处插入到公园之中，慢慢地逐渐向乡村区域延伸与其融为一体。在此项目中，自然和人工景观之间的分界线显得更为明显，通过一个5米的转变层将已完工区域分为两半。乡村和城市之间的物理屏障阻碍了我们对自然的探索。其结果导致了一种临界边情况，即两种对比鲜明的现实情况之间的界限，不断地寻求着在双方之间进行对话的可能。

这两座建筑形成了两座教学大楼，它们是该座城市在Agro Romano的理想延伸，是从城市活动到自然世界的扩展。它们是居民中心区的逐步扩展和乡村的逐步缩减之间的视觉桥梁。人工环境扩展入乡村区域，自然环境又渗透其中，形成一个不断重叠的过程。其外部空间相互交织着进入学校区域，又依次延伸至其周围环境中。这些建筑物开放至其环境之中，在内部和外部之间形成持续的对话和压力。通过延伸至公园之中，这些建筑与自然融为了一体，不再是封闭型的街道表层结构，而是向其周围环境逐渐开放，而持续插入其中的自然环境也缓慢地变成了建筑结构的一部分。

而这些建筑物内部空间的联接形式又反射出了其转变过程，其向外部环境开放和关闭的过程。为了同教室内部的免费而有组织的活动空间相匹配，此建筑的闭拢高度为1.2米，形成一个大型光滑的水平切面，使其能够同外部环境永久保持视觉联系。而与此相反，其工作室内空间的一侧为封闭式墙，只有一些小的水平开口，还有一个全高度光滑表面，可将其内部空间投射到罗马城四周的平原上，形成了观赏此处自然景观的一个特别视点。

该公园与沿着本项目的整个边缘地区分布的板式结构一同设计。此板式结构具有交互分布的休闲区以及用于提供阴凉的自然植被区，形成了其自身的特色。这正是那条穿过公园和项目居住区进入Agro Romano自然景观区的贯穿轴起始点。

从这些通道进入到公园之中后，逐渐失去了其连贯性，完全摒弃了其物质形态，变成了具有简单标志的古旧土路。而采用自然材料制成的另一个通道网络，同前一连接系统一同贯穿于整个公园。其组成成分根据不同的活动而各不相同：分别包括自行车道、步行道、运动跑道和教育通道。

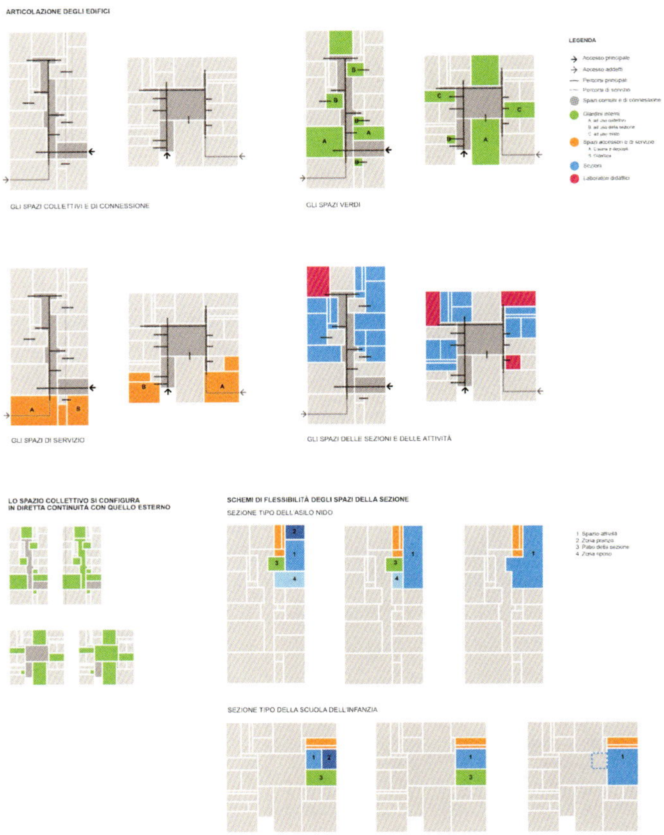

New School Rome

新罗马学校

International Shortlisted Competition "Tre Nuove Scuole a Roma" – Casal Monastero - Second Phase | Design: n!studio Susanna Ferrini, Antonello Stella | Team: Simone Bove, Giulia Scaglietta, Gabriele Misso, Loris Rossi, Rocco Smaldone | Collaborators: Simone Piacenti | Engineering: Pr.as. Progettisti Associati, Ing. Massimo Calda | Consultants: Carolina De Camillis, Riccardo Fibbi Lighting Desig | Prof. Dario De Luca Educational Expert, Umberto Pessolano Geologist | Client: Rome Municipality | Period: 2005 | Model: Marco Galofaro

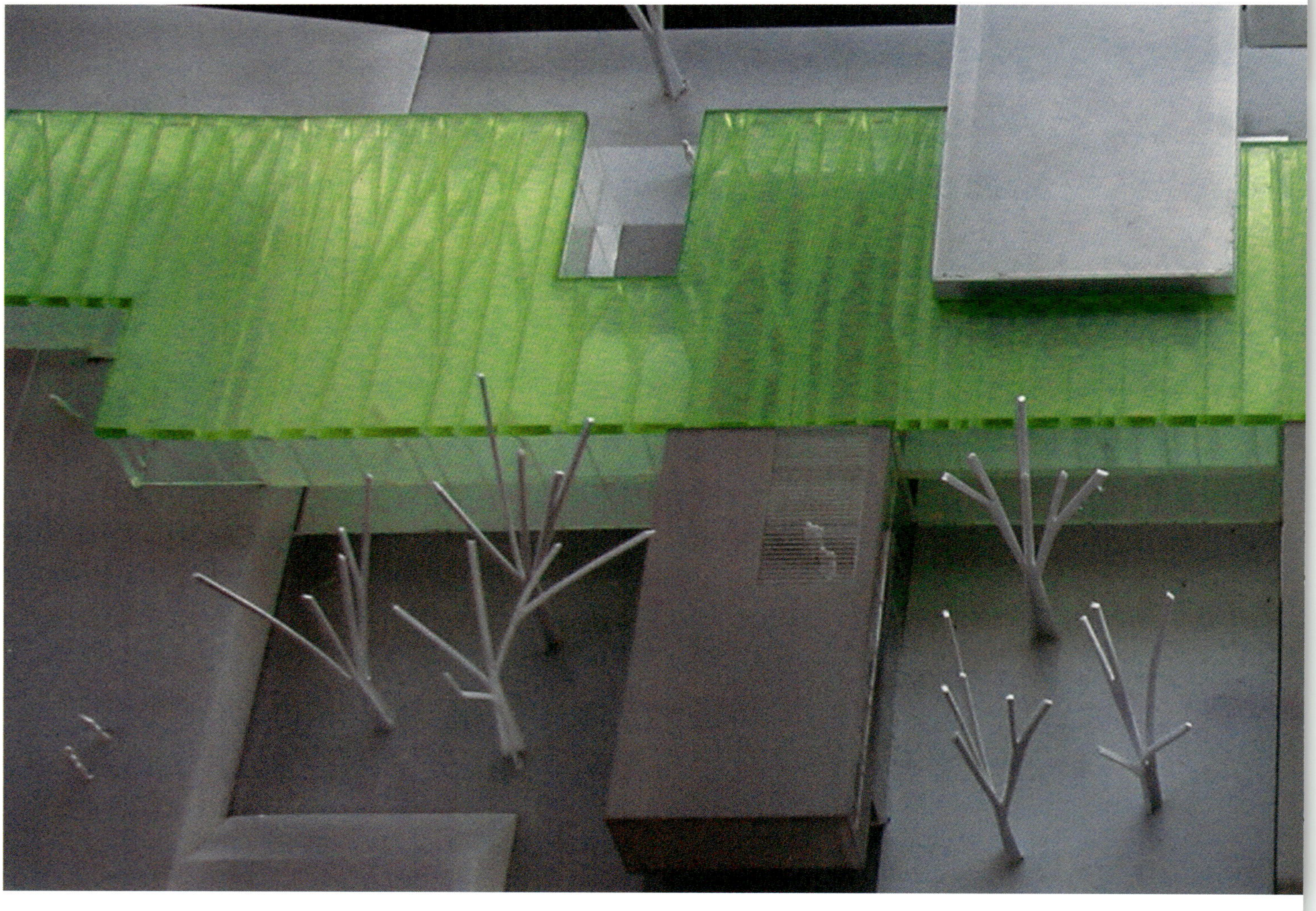

sense of matter

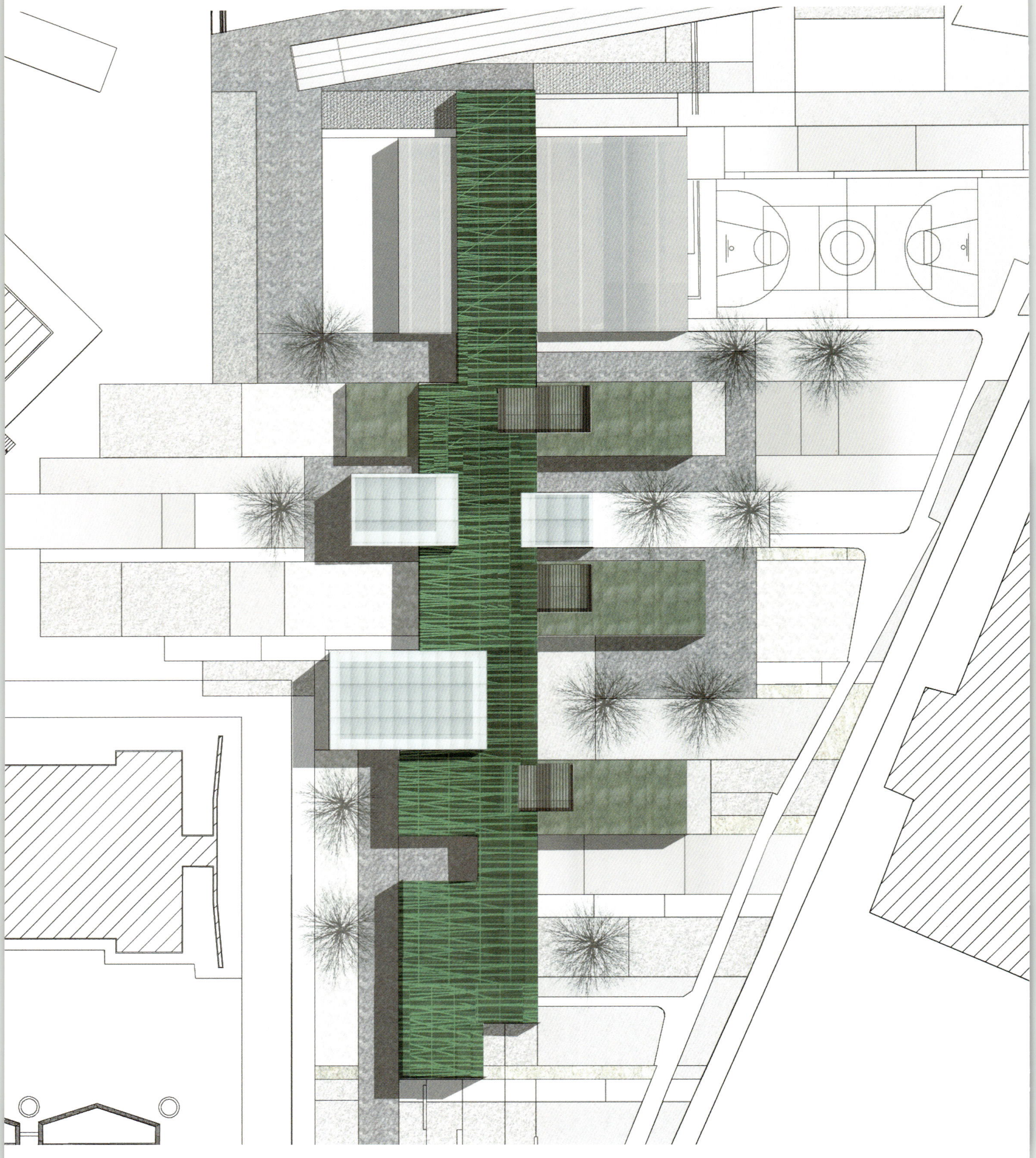

The theme of the competition was the construction of a new high school in the Roman periphery of Casal Monastero, an area dominated by linear residential buildings and other scholastic complexes. The premise that lies at the base of our project was that of considering the school as an organism that is open towards the extended space of the neighbourhood. This desire for a strong level of interaction between public and scholastic space leads almost 'directly' to the form of the school building: a central path, the ideal extension of the path that winds through the neighbourhood to the scholastic centre, crosses the entire building on two levels: the upper at the level of the neighbourhood and the lower at the level of the park. This premise thus generated, as the first act of design, the volumetric solution adopted, which renounces the use of a compact and monolithic volume, in favour of an open solution where equal importance is given to the spaces of the school and the landscaping. The school occupies two levels and is based on an axial path of internal circulation that begins in the entrance hall and 'expands' along a single-loaded corridor towards the classrooms and other public spaces in the school: the aula magna, the library and the music room. This space of distribution terminates in the cafeteria and gymnasium, spaces that overlap one another at different heights: from the cafeteria it is possible to look down into the space of the gymnasium and vice versa. The unifying element is that of the roof, imagined in a composite material of light green pigmented fibreglass with bas-relief cuts on the underside, reproducing, in a stylised manner, an ideal pattern of branches and underlining the integration between the building and the surrounding landscape.

本次赛事的主题是在罗马周边的Casal Monastero区建造一所新的高级中学，该区域是一个主要由直线形住宅建筑和其他学校综合建筑组成的区域。项目的前提是将学校看作是一个有机体，并向邻近延伸区域开放。需要在公众区和学校之间建立高度互动关系的要求几乎“直接”导致了该学校的建筑模式的形成：即修建一条中心通道，通道的理想延伸模式就是在邻近区域盘绕而行，而后抵达学校的中心。它在两个层次穿过整个建筑：上层穿过了邻近区域，下层穿过了一座公园。设计与这种需求相匹配，它采用了一种整体解决方案，摒弃了紧凑型整体式结构，支持一种同时重视学校空间和园林风景的方案。该学校占据两层空间，以一条内部循环通道轴为基础，该通道从入口处开始，沿一条单边走廊“延伸”到教室及学校中的其他公共区域：学校礼堂、图书馆和音乐室。这种空间的分布到自助餐厅和体育馆结束，在不同高度不同空间相互交迭：从自助餐厅向下可以看到体育馆的空间，反之亦然。将二者统一到一起的结构就是其屋顶，该屋顶由一种底侧具有浅浮雕雕刻且着绿色的玻璃纤维合成材料构成，以一种固定格式的方式复制了理想的分支类型，使该建筑和周围风景相结合。

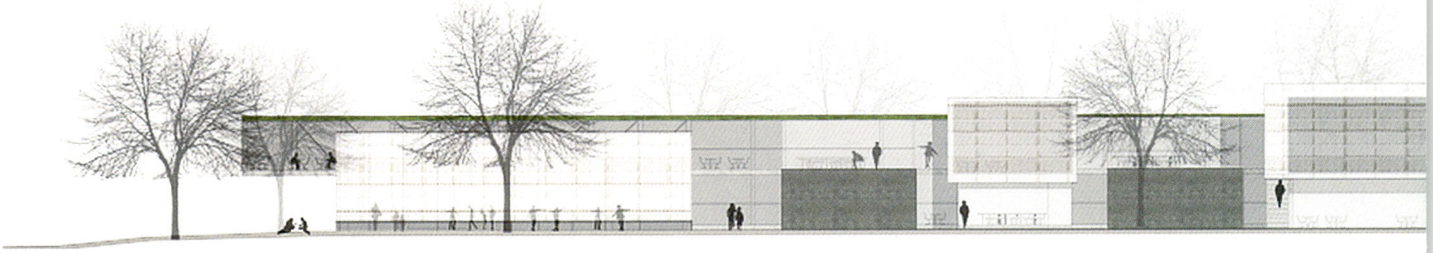

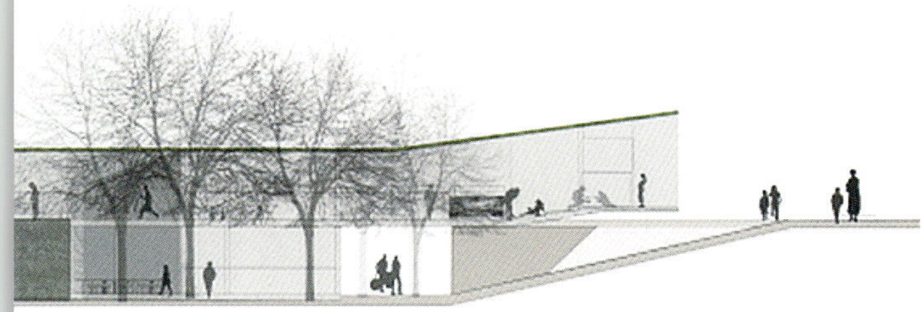

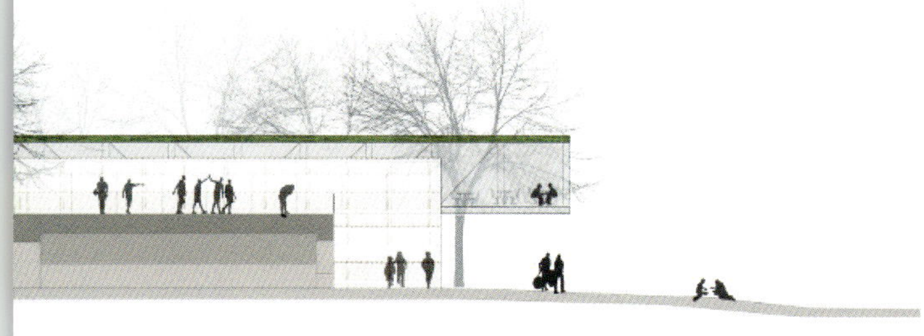

AULA MAGNA

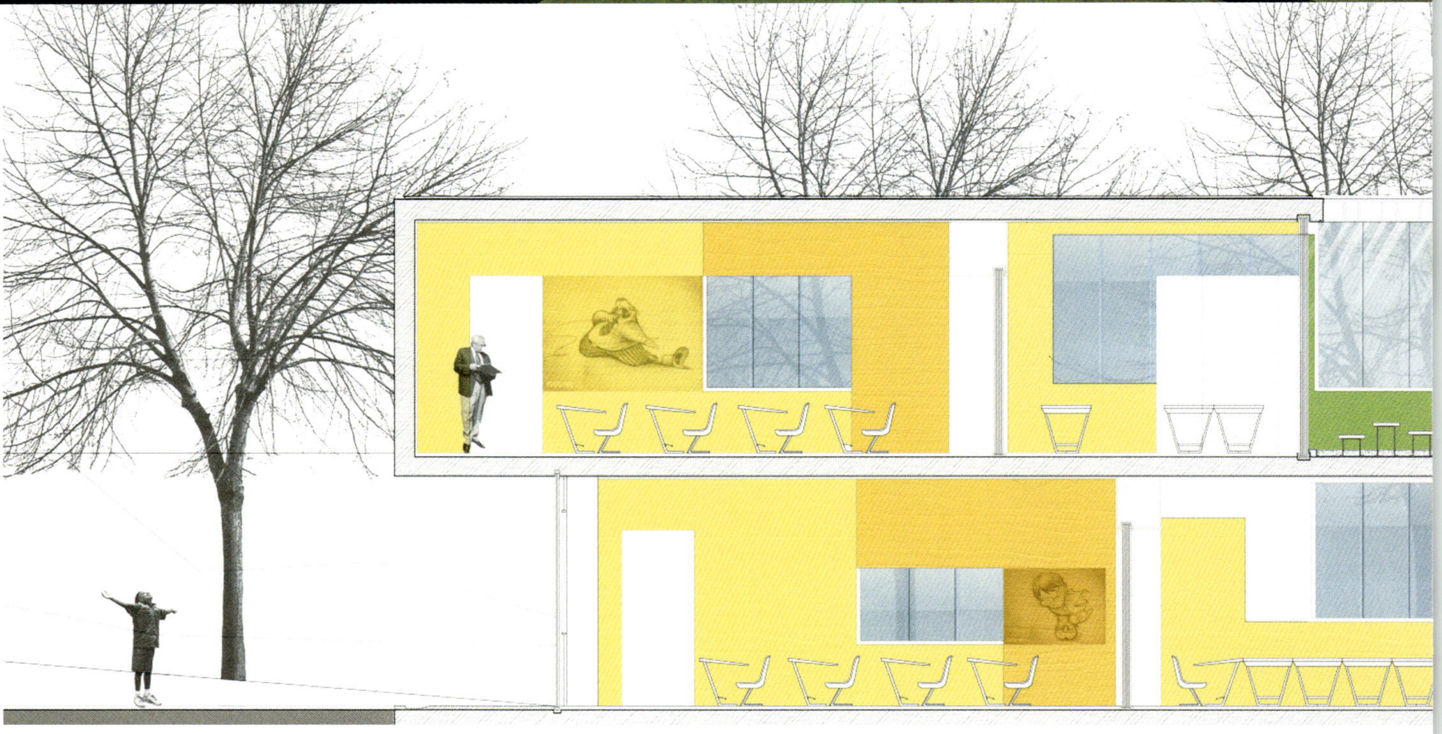

Temporary Storage Structure

临时性储藏建筑

Temporary Structure for Archaeological Remains inside the 'Villa dei Quintili', Rome | Design: n!studio Susanna Ferrini, Antonello Stella with Luigi Filetici
Team: Carlo Celia, Piero Neogrossi | Coordinator: Maria Grazia Filetici | Engineering: Prof. Ing. Salvatore D'agostino | Collaborator: Claudia Farinelli, Jutta Ri
Client: Sovrintendenza Archeologica di Roma, Dott. Rita Paris Contractor: Officina Ragnini Pitigliano | Period: 2001-2002 | Award: Monitorp, Bauwelt 2005,
Architectural Review Award 2006

sense of matter

The building is located inside the villa dei quintili archaeological site on the via appia antica, a true example of a public archaeological park. This temporary building, for the storage, cleaning and restoration of archaeological remains, represents an experiment in the insertion of the 'new' within such a delicate and important historical and natural context. The nearby roman cistern, the surrounding vegetation and the relationship with the landscape within which the building is visually inserted, the very 'ground plane' upon which it rests, have all, in some way, determined the choices related to the building materials, in this case essentially two: glass and cor-ten steel. Large sheets of glass, without any external frames, create the main façade that, based on various conditions of lighting, either reflects the landscape or contributes to an image of total transparency. Cor-ten steel, used in various configurations to create the 'shell' of the building, as a result of a natural process of oxidation, tends to blend in with the colours of the natural and historical context. The temporary nature of this 'lightweight' steel pavilion is underlined by its 'suspension', the very act of resting temporarily on the ground. The long sides of the building are treated in a different manner: the design of the cor-ten panels on the entry façade represents the 'stacking' of the containers of archaeological remains inside the building; the other, necessarily more closed, employs fragmentation to divide the surface into numerous, thin parallel strips of 'u' profiles in cor-ten steel, reducing the compact nature of the volume, which appears to be both low and long, similar to the nearby roman cistern.

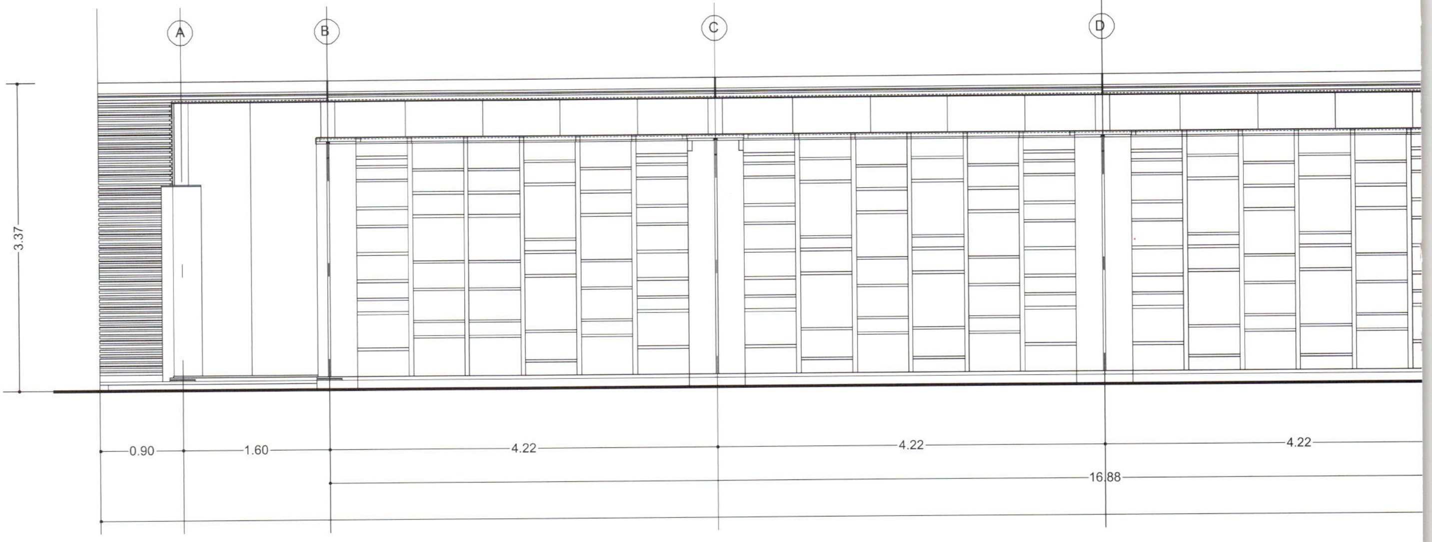
A
B
C
D
3.37
0.90
1.60
4.22
4.22
4.22
16.88

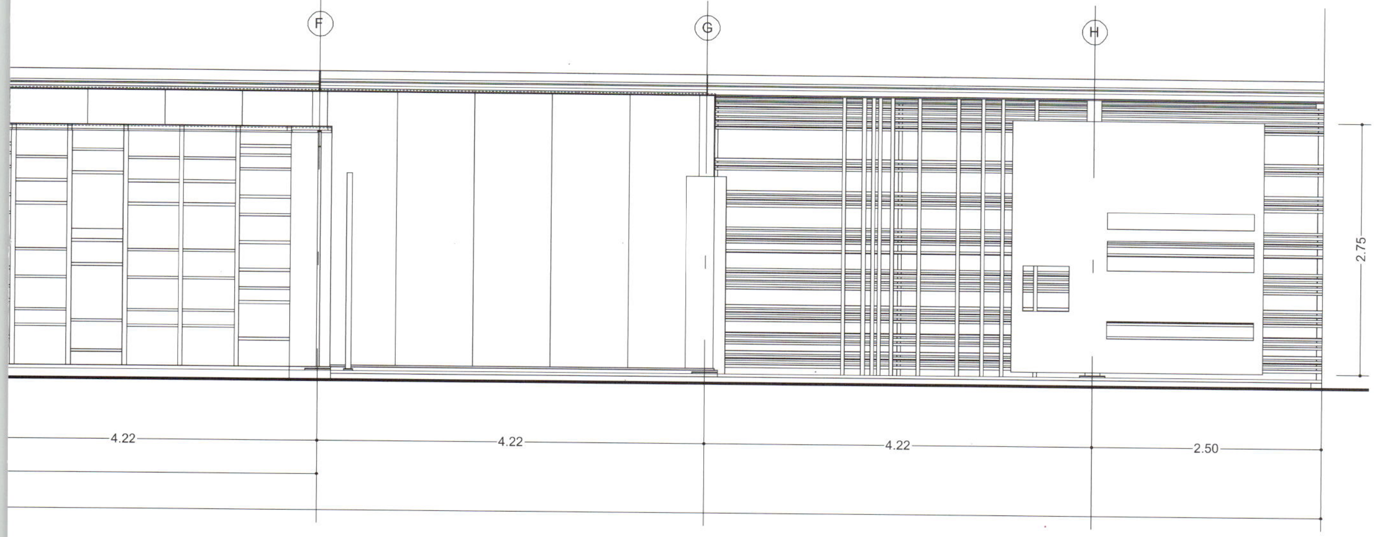
F
G
H
2.75
4.22
4.22
4.22
2.50

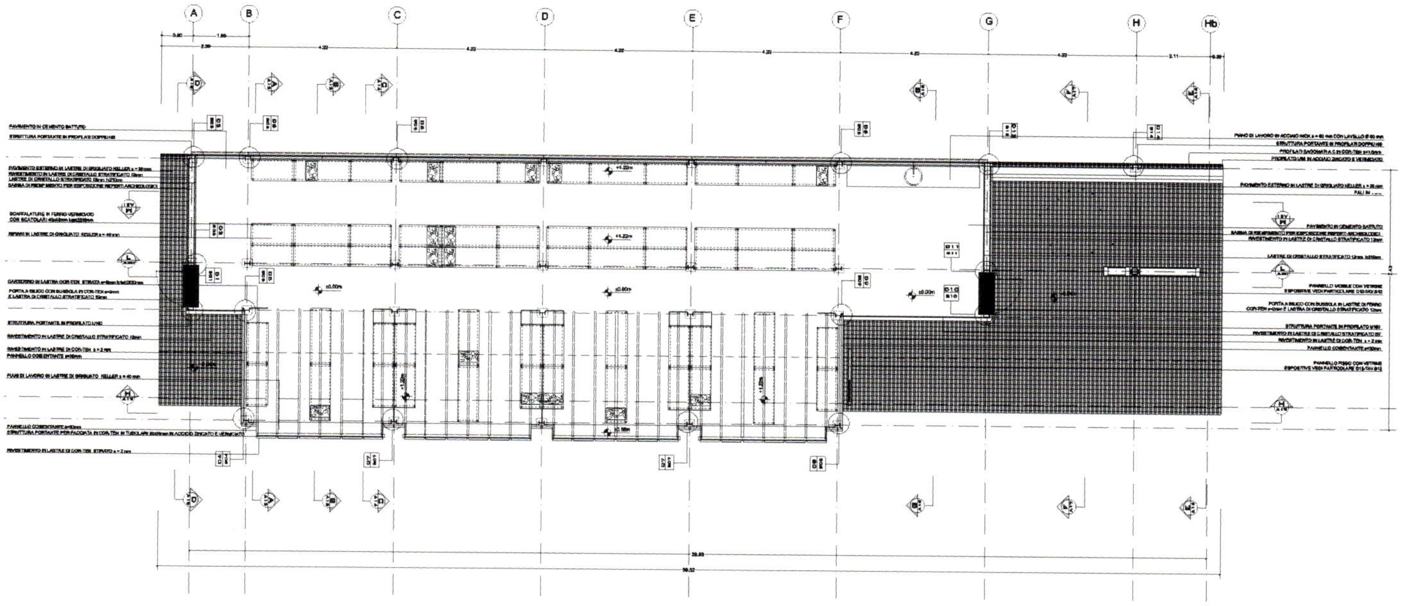
A
B
C
D
E
F
G
H
Hb

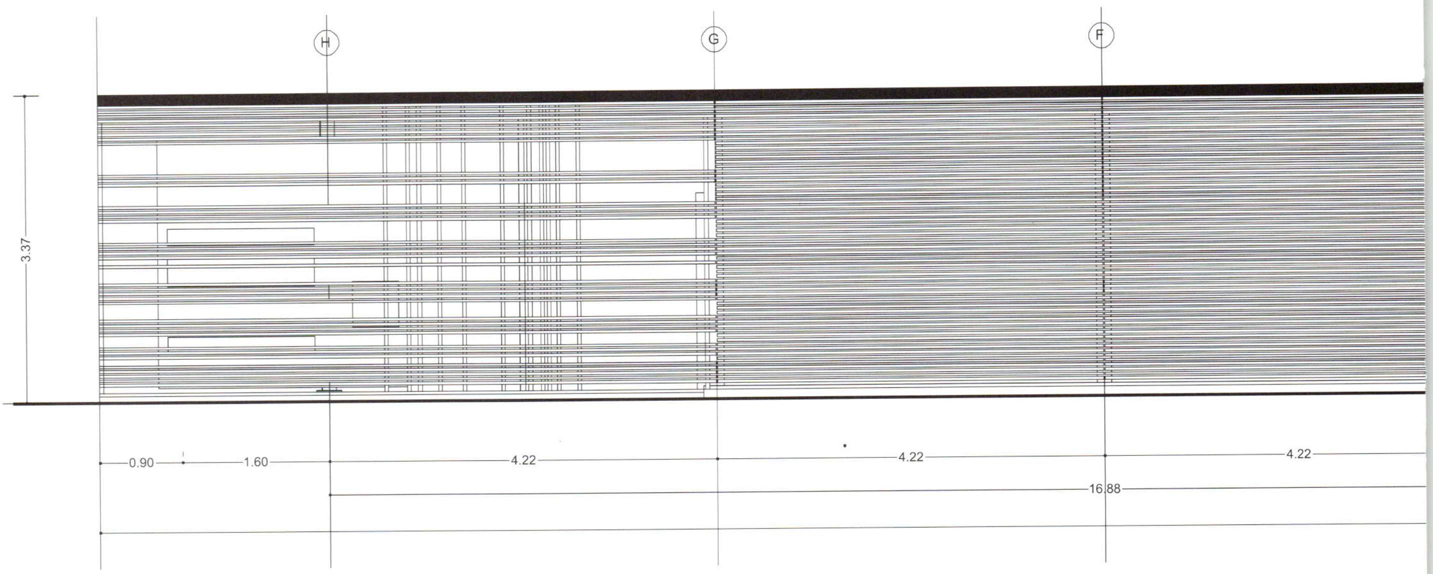
H
G
F
3.37
0.90
1.60
4.22
4.22
4.22
16.88

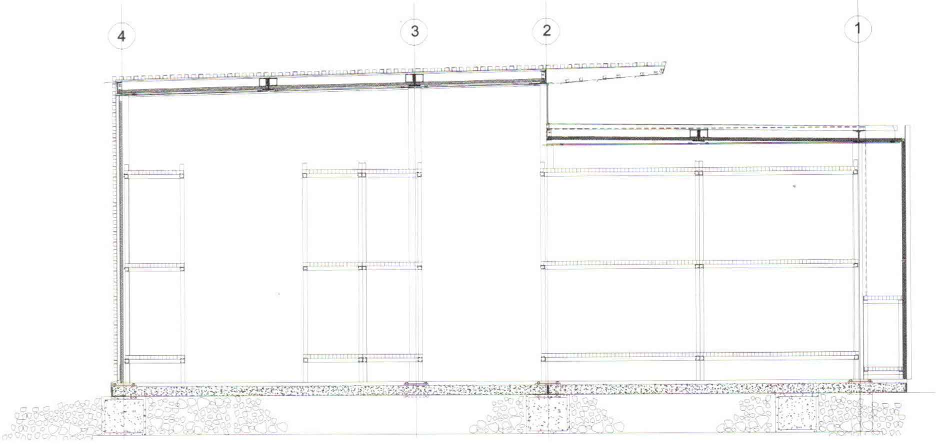
4
3
2
1

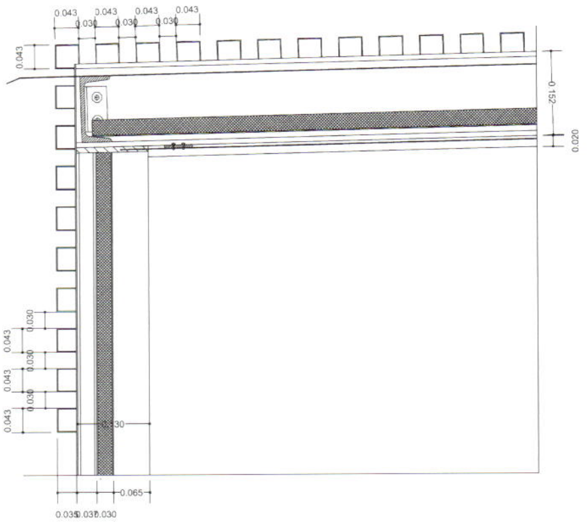

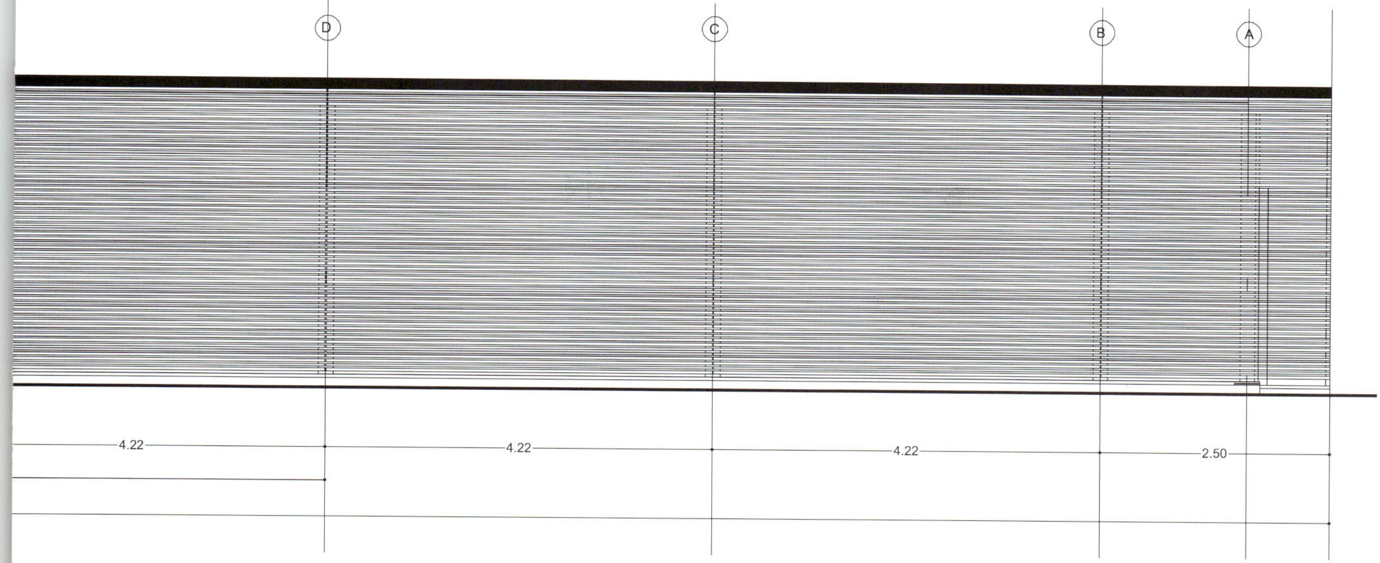

此建筑位于Via Appia Antica的Villa dei Quintili考古区内部，是一个公共考古公园的真实实例。这座用于储藏、清洁和修复考古遗迹的临时性建筑是将“新建筑”插入到非常易损且重要的历史和自然背景环境中的一种试验。附近的罗马蓄水池、周围的植被及该建筑与其所插入的视野范围内的风景之间的关系，都在某种程度上决定了其建筑材料的选择。具体对本建筑而言共采用了两种重要材料：玻璃和低合金高强钢。没有任何外部框架的大片玻璃板构成了其主要的正面外表面肌理，根据各种各样的照明条件，反映出了其独特的外立面，增强了其全透明的形象。可用于各种各样构造中的低合金高强钢构成了该建筑的“壳”，经过自然氧化过程之后，它将与周围的自然和历史环境融合到一起。通过其“悬空性”更加突出了此“轻重量级”钢质亭阁的临时性特性，它刚好是临时停留在地面上。而此建筑较长的一侧则采用了一种不同的处理方式：入口正面外表面的低合金高强钢板代表了建筑内部考古遗迹容器的“层叠错落”，而不可避免地需要采用更加封闭的形式；另一面则通过分割的形式将表面分成了采用低合金高强钢板构成的众多狭长的平行U形带结构，降低了此建筑的紧凑性，使其显得既低又长，同附近的罗马蓄水池相似。

sense of matter

Mediterranean Museum Conservation Centre

地中海博物馆保存中心

ernational Shortlisted Competition for the Former De Muy Barracks in Marseille - Second Phase | Design: n!studio Susanna Ferrini, Antonello ella (Coordinators) with Giovanni Bulian, Amedeo Schiattarella, Nicoletta Trasi | Team: Simone Bove, Gabriele Misso, Giulia Scaglietta, Andrea hiattarella, Francesco Sabatini, Alessandro Turina | Consultants: Carolina De Camillis, Riccardo Fibbi Lighting Design | Engineering: Beterem genierie Sa, Ing. Luc Delasalle | Collaborators: Rocco Smaldone, Paolo Soellner, Bertrand Lascourreges | Client: Ministere De La Culture Et De La mmunication, Direction Des Musees De France, Emoc | Period: 2004 | Model: Marco Galofaro | Prize: 3rd Place

The restricted procedure competition programme called for the transformation of the area of the muy barracks in marseille into a centre for the storage and restoration of the objects to be displayed in the new museum of european and mediterranean civilisation. The project is developed around two, parallel themes: on the one hand the necessity of guaranteeing differentiated paths and controlled access for visitors and researchers, on the other the desire to model the ground plane, opening up the areas of the storages towards a public park to be built in the area. The objective of the project is that of activating a new relationship between the neighbourhood and the conservation centre and proposing a new idea of storage building, introducing spaces for public visitation and defining the ateliers as a focal point within the project. The conservation centre is thus no longer conceived of as a traditional warehouse, closed and invisible but, on the contrary, becomes an organism that is open towards the city, a complex that can be perceived and placed on display as a container of the culture and civilisations of the mediterranean: "the mediterranean and the discussion of the mediterranean are inseparable". (Predrag matvejeviç) our desire was in conflict with the programmatic request of the competition brief, which sought the almost total occupation of the site; the strength of our project was that of responding to the requests of the programme, while freeing up the largest possible area, leading to the idea of breaking up the complex into various volumes and basement spaces, inserted within a public urban park.

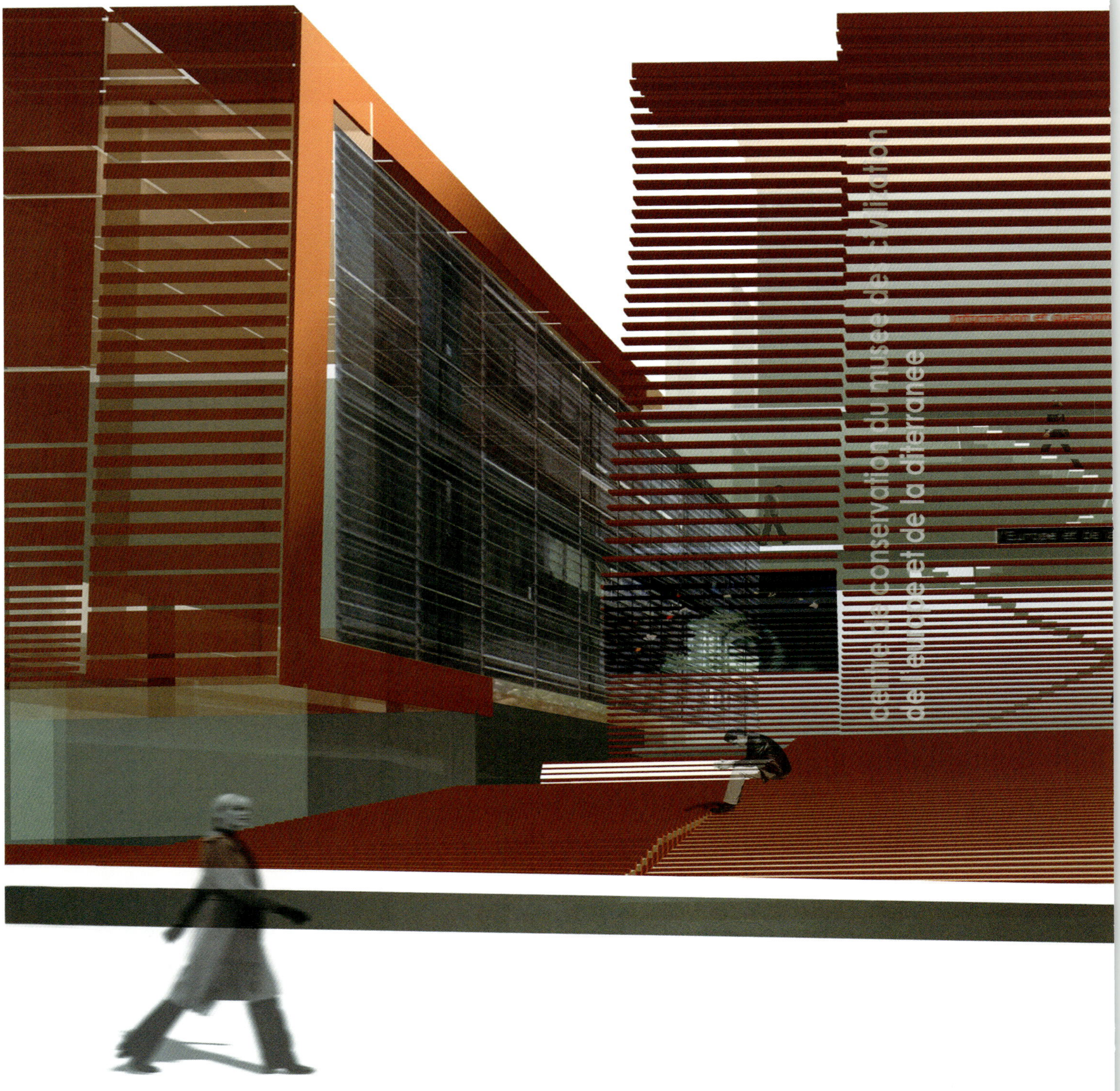
de conservation du musée de
de l'europe et de la

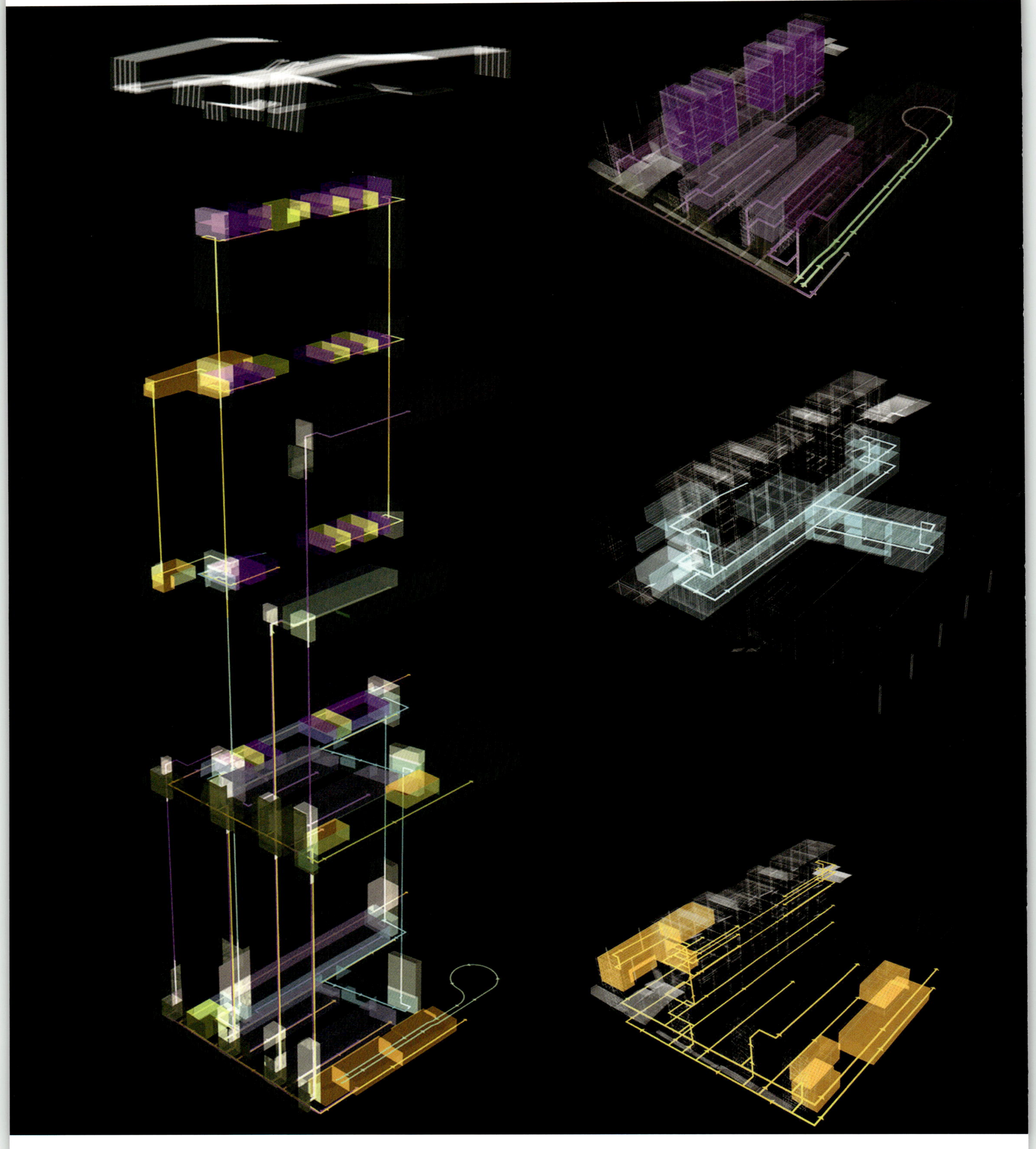

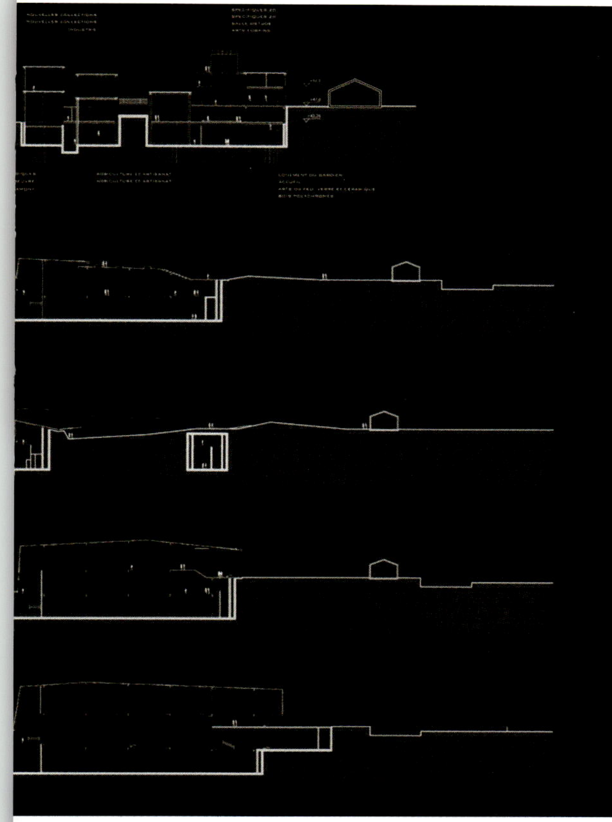

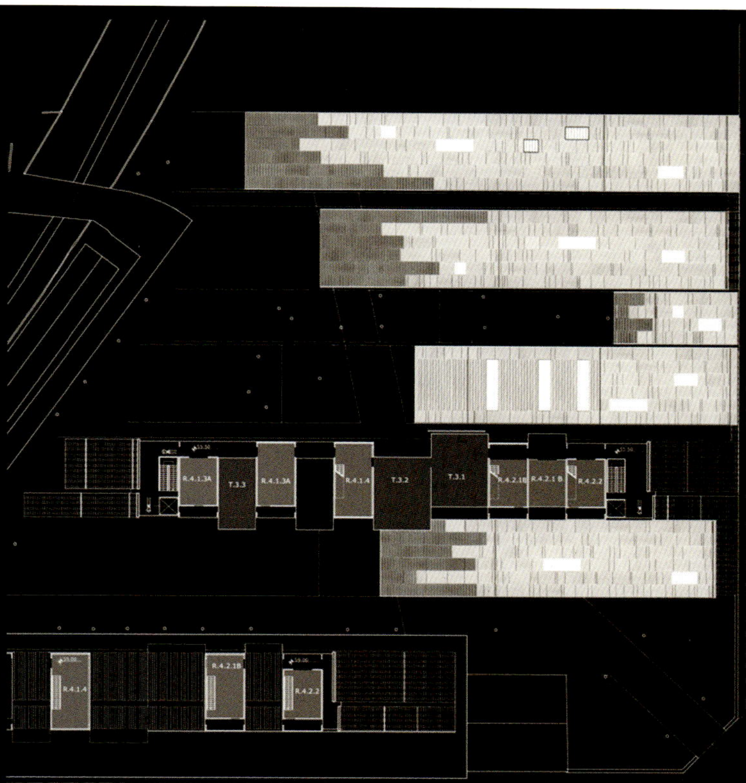

R.4.1.3A
T.3.3
R.4.1.3A
R.4.1.4
T.3.2
T.3.1
R.4.2.1B
R.4.2.1 B
R.4.2.2
R.4.1.4
R.4.2.1B
R.4.2.2

此竞赛项目要求将Marseille的Muy兵营区改造成新的欧洲和地中海文明博物馆展品的储藏和修复中心。此项目沿着两个平行的主题展开：一方面需要保证游客和研究人员路径各自分开并有不同的进入权限；另一方面希望对其地面进行设计，将其储藏区向将在该区域建造的公园开放。该建设项目的目的是在周围区域和该保存中心之间营造一种新的关系，提出一种新型储藏建筑理念，增加公众游览空间并使工作室成为该项目的核心点。这样人们就不会再认为该保存中心仅是一个传统意义上的仓库，封闭而隐蔽。恰恰相反，它将变成一个向城市开放的机构，一个能够作为地中海文化和文明的载体而被认识、被展示的综合体："地中海和针对地中海的讨论互不可分"（Predrag Matvejevic）。此竞赛章程要求充分利用该地区的所有面积，但我们的期望同它的程序要求有些冲突。我们的项目的优势在于在响应该程序要求的同时，还可以解放出大量的空间，从而得出了将该复合建筑分割成不同的建筑体和将地下空间分别插入到城市公园中的想法。

Ecoromabuilding

生态罗马建筑

Design for a Sustainable Residential Building in Vigna Murata, Rome | Design: n!studio Susanna Ferrini, Antonello Stella | Team: Simone Bove, Stefano Battaglia, Anna Hollstein, Angelo Venturi | Client: Parsitalia Rea I Estate | Period: 2007 | Model: Simone Bove, Anna Hollstein n!studio

FAR
CLOSE

The design of the Vigna Murata Tower is based on four themes: sustainability, new lifestyles, a relationship with the landscape and contemporary ideas of Romanity.
Located in the southern sector of the Capital, adjacent to the EUR business park, the project site is part of an Integrated Programme for the "Città da Ristrutturare" [The City to be Renovated] focused on the improvement of urban quality and, in particular, the modernisation and integration of mobility and services through private competitions.
The new urban structure, defined by the integrated programme, calls for a new road axis and the functional mix of all new constructions - residences, services and commercial spaces - all of which is immersed in an urban park structured by a network of pedestrian and bicycle paths and facilities.
The objective of energy savings and the environmental impact of the new structure were accompanied by a particular focus on the landscape and the relationships created by the intervention. The site is located adjacent to one of the many fragments of parkland - true remnants of the Roman campagna - that mark and structure the city of Rome in a particular way.
The integration of the landscape, the search for a re-proposal, in a new and contemporary manner, of the relationship between the building and its context has marked, in methodological terms, the traces, orientation and directions of the project's development.

本项目的场地位于意大利首都的南部，邻近EUR商业园，此项目属于"Città da Ristrutturare"(需要翻新的城市)这一综合项目的一部分。项目建设的目的在于提高城市质量，尤其是提高其现代化水平和通过私有竞争融合移动性和服务性等方面。
此综合项目所界定的新型城市结构要求建造一个新的公路轴线以及所有新建筑的功能组合——包括住宅、服务和商业空间——所有这些均融入一个由新的人行及自行车通道和设施网络构成的城市公园中。
在节约能源和降低新建筑结构对环境影响的同时，此项目尤其注重其地貌风景与新介入建筑所形成的关系。该项目的场地位于众多公园的旁边，这些园地是罗马周围平原的历史遗迹，它们以一种独具特色的模式对罗马市进行了点缀、规划。
其在地貌风景方面的融合作用，是以一种新型、当代的模式重新寻找建筑物及其背景之间关系的过程，以一种方法学的模式标示出了此项目发展的痕迹、定位和方向。其所作出的选择就是通过利用一种特殊的关系与当地地貌风景关联起来，现在已经成为了罗马市郊统一且非常重要的一部分，在乡村地区和传统古代及中世纪遗迹之间保持了一种牢固的联系，它是自然和人工、风景和建筑之间非凡融合的图形信号。
此座同时为当地居民提供公共服务的住宅建筑高38米多，以将其对地面环境的影响尽可能降低。一个巨大的开口穿过此建筑，并连通了两个花园，而此建筑正好插入这两个花园中间，其绿树葱葱的凉廊系统环绕在该建筑物周围，将自然景致融入到建筑的外结构之中。

SCHOOL
PARK
ECOBUILDING

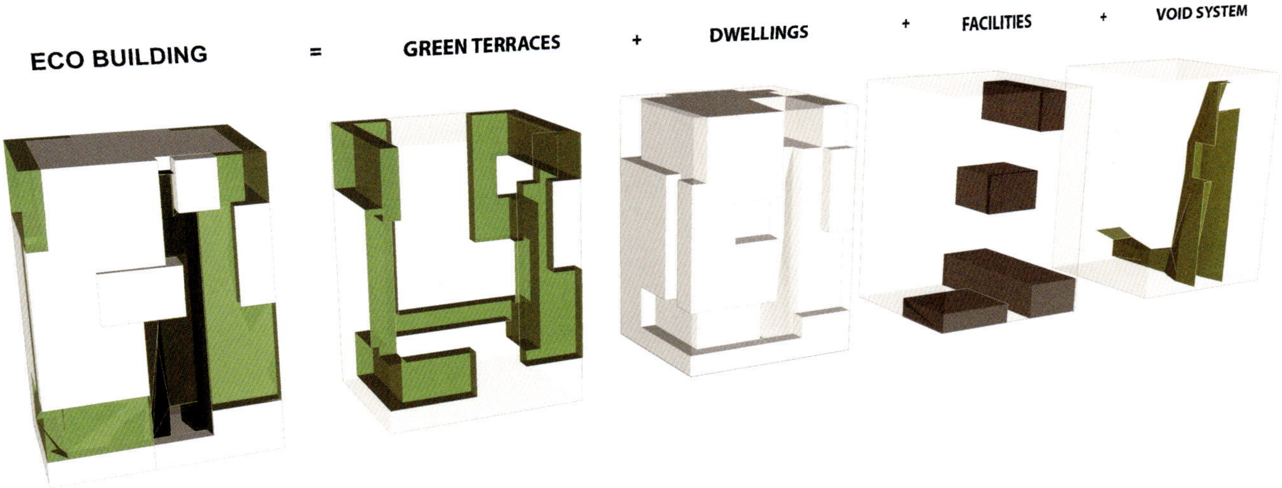

Schematic plants and facades

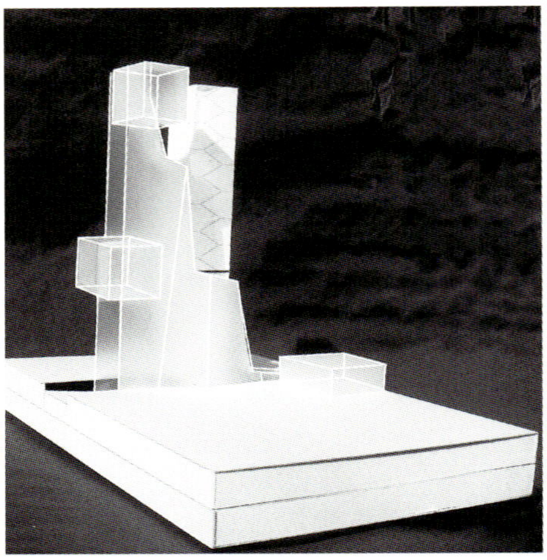

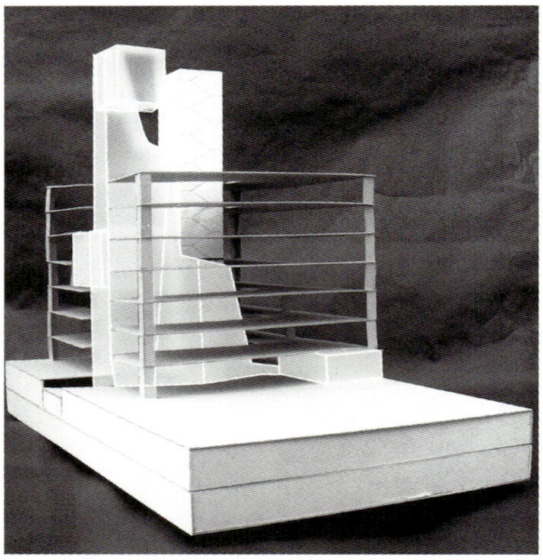
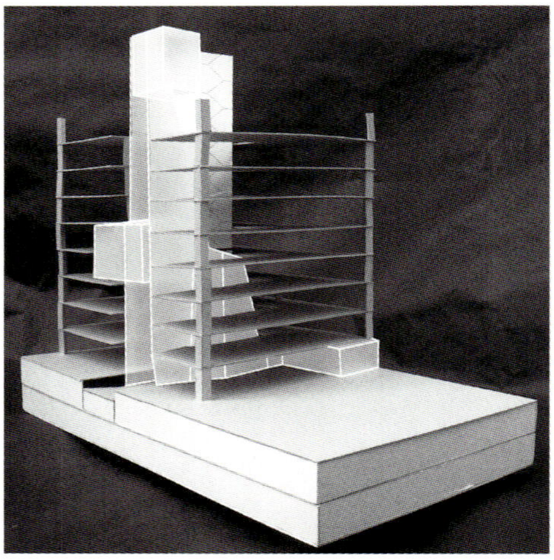

The choice was that of relating to the landscape by drawing on the special relationship, now a consolidated and fundamental part of the Roman suburbs, which maintains a solid connection between the countryside and the ruins of classical antiquity and the medieval period, the figure-symbol of an extraordinary integration between nature and artifice, between landscape and architecture.

The residential building, which also features public services for local residents, is over 38 metres in height, in order to reduce its impact on the ground, as much as possible. A large cut allows it to be crossed, creating a connection between the two gardens into which the building has been inserted, while a system of planted loggias wraps around the building and integrates the natural element of the façade.

The cut that splits the building and the green terraces on the façade contribute to improving the thermal performance of the building during the winter and summer, welcoming the surrounding landscape of the Roman countryside. Apartments of various typologies and dimensions spin along the height of the building in close connection with the terraces in a search for optimum lighting, even in kitchens and bathrooms. Public areas for collective activities are placed along the cut, establishing a direct relationship with the surrounding landscape. The building is thus perceived as a carved volume with strong chiaroscuro effects, similar to the Roman ruins that dot the city's urban and sub-urban landscape. This effect is strengthened by the design of the façade, which makes use of different layers of perforated burnished brass panels that modify the perception of the building based on the distance from which it is observed: seen from afar it recalls the Roman ruin; from up close, it reveals its brick-like texture. Seen from the proper conceptual distance, the project represents our idea of Romanity, rendered through perception rather than shape.

将此建筑分割开来的开口以及外表面结构绿色的阶地有利于提高该建筑在冬天和夏天的保温性能，与周围罗马乡村的独特景致相适应。各种类型和大小的公寓空间螺旋分布在大楼外侧直至楼顶，同阶地紧紧相连以获得最佳的采光效果，即使是厨房和浴室也是如此。用于举行集体活动的公共区域沿着其开口分布，与其周围的风景建立起了直接的联系。这样该建筑物就可以被当作是一座具有强烈明暗配合效果的具有雕刻感的建筑体，同散布在罗马市和罗马市郊风景中的罗马废墟一样。其外表面结构的设计进一步加强了这种效果，该设计采用了不同层的带孔抛光铜板，这样根据距离的不同人们对该建筑的感觉也会发生改变：从远处看，它使人们回想起了罗马废墟；在近处看，它则会显露出其砖样结构。在适当的概念距离观察，此项目能够代表我们对罗马的看法，通过感知而不是形状的方式展现在我们面前。

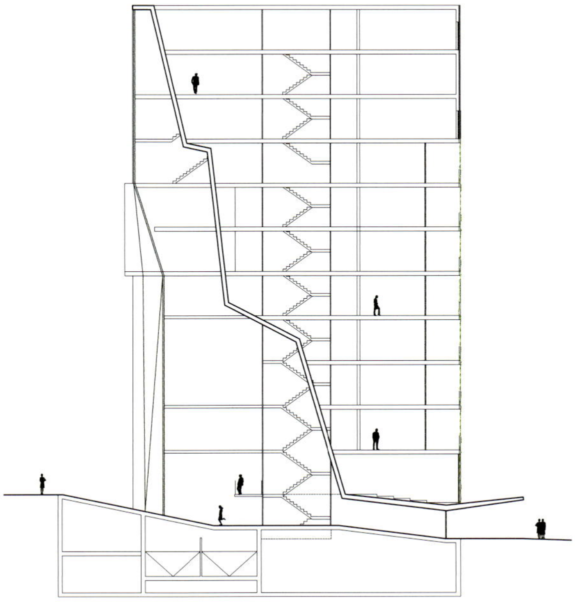

East Gate Tower

东门塔楼

Mixed-used Tower, Padova | Design: n!studio Susanna Ferrini, Antonello Stella, with Andrea Pasquato Architects | Team: Stefano Battaglia, Simone Bove, Anna Hollstein, Angelo Venturi, Paolo Rossi, Fabrizio Romano, Lara Turchini | Client: Enrico and Giorgio Scattolin | Period: 2008

This project aims to give the city of Noventa Padovana a clear and tangible "gate", composed of a new architectural element that marks the entrance to the city and is characterised by an elevated level of environmental and mediatic quality; a landmark that contains highly representative functions: an element at the "urban scale" that constitutes the access portal evoked by the project slogan: East Gate. The creation of the new high-speed connecting road offers opportunities for clearly and effectively marking the new points of entry and crossing of this coastal city, while simultaneously weaving the fabric that runs between this new architectural portal and the accidental constructions found in the municipal territory. This is obtained by inserting a building that "communicates" using the language of the site, loaded with dynamism and a shifting and chameleon-like appearance. While the new building is undoubtedly technological, a clear reference to the world of media, it respects the highest levels of environmental sustainability. It is an architectural object that dialogues with the natural world. In reality, it assimilates it in its folds, creating hanging gardens at multiple levels, and using nature to define the facades and characterise the spaces of the wellness/fitness centre located on the top floor of the horizontal block. The north façade is screened by a technological surface that, based on the project objectives, assumes the appearance and

function of an artificial and mediatic "digital skin".
This screen is dotted with glowing elements that transform the entire structure into a media-wall that s visible from a great distance, primarily from the high-speed arterial roads, such as the autostrda and the ring road.
This apparatus exalts and amplifies the intentionally dynamic qualities of the architecture of infrastructural and traffic flows that surround and lap at the edges of the project site.
The south façade of the tower opens towards the surrounding territory, triggering a strong dialogue with the landscape. In fact, this façade is much more permeable and characterised by the important presence of insertions of nature and vast openings towards the exterior. The use of natural elements as an architectural building block and connection with urban spaces has a significant impact on the "mediation" between the building and the heterogeneous residential areas behind it. The particular sloping geometry of the urban park to the east of the poly-functional building is used to meditate and connect the two contexts described above through the introduction of a "dialectic" focused on creating a dialogue and compromise, in which distance are bridged by a vast urban park with various public facilities.

此项目的旨在为Noventa Padovana市建造一个明确、真实的“大门”，可作为该城市入口标志的新型建筑物，它具有高度环保和可调节的特质，此标志性建筑构成了城市的入口，正如该项目的宣传口号所述：东大门。

建成的新型高速公路成为进出此海滨城市的新起点，同时将此新建筑和其他各个建筑之间的网络结构交织在一起。

这是通过插入一种具有动态、变化、变色龙样的外表，可通过场地语言进行“沟通”的建筑来实现的。此新建筑毫无疑问非常具有技术含量，是这个媒体世界的明确典范，同时还符合最高等级的环保要求。它是能够同自然世界进行对话的建筑体。事实上，它采取了一种多层相似层次的模式，在许多层均建造了悬空式花园，利用自然背景来界定其正面结构，在卧式模块的顶层形成了极具特色的健身中心。其正面和背面由电子显示屏构成，具有人工和媒体性“数字表层”的外观和功能。

屏幕上点缀着闪光元件，可将整个建筑转变成一个巨大的媒体墙，可以在Autostrda环路和其他直通高速公路上远距离观察到。此装置提升了此基础建筑设施的预期动态质量及项目场地周围及边缘区域的交通流量。

该塔的正南面面向周围的区域，同地貌风景之间形成了强烈的对话。事实上，此结构的可透过性很强，其外向的天然巨大开口形成了建筑的重要特色。将天然元素用作建筑模块，以及同城市空间之间的联系，对该建筑物和位于其后面的各种各样的住宅区之间的“调和作用”具有非常重要的影响。位于此多功能建筑东面的城市公园是一个具有特定坡度的几何形状，通过引入一种“辩证”的结构来集中营造出一种对话和调和的结构，建立具有各种公共设施的宽广的城市公园的桥梁作用。

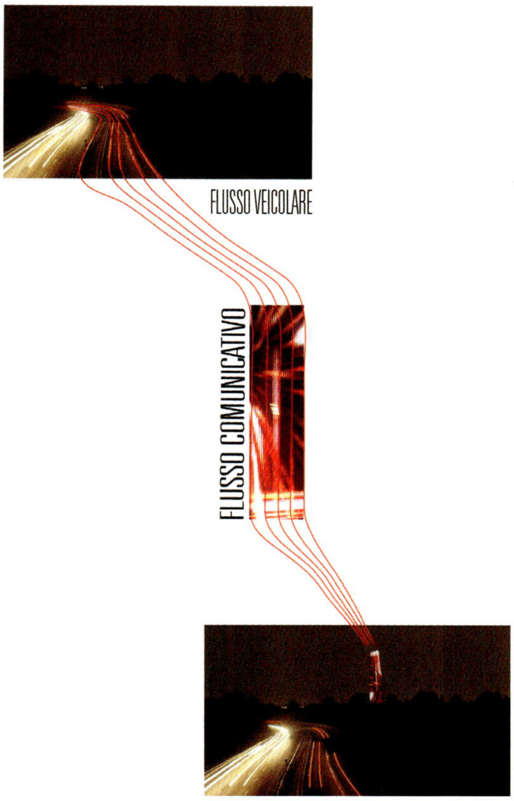
FLUSSO VEICOLARE
FLUSSO COMUNICATIVO

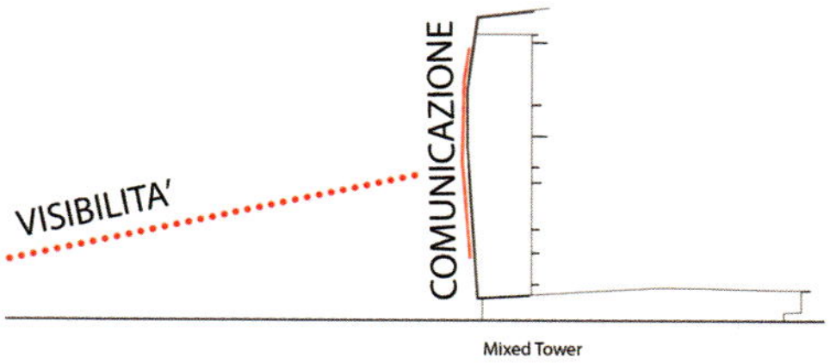
VISIBILITA'
COMUNICAZIONE
Mixed Tower

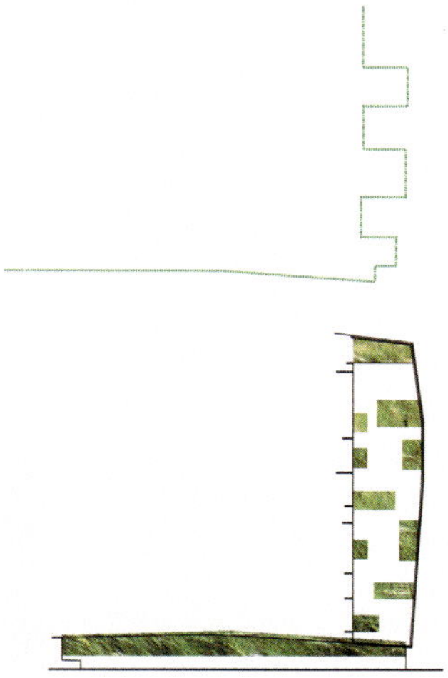

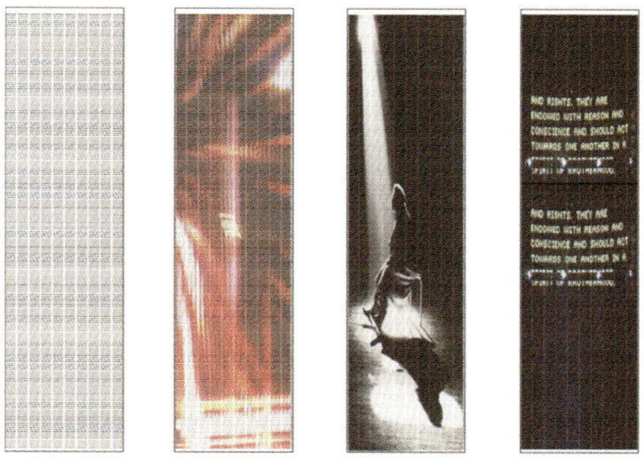

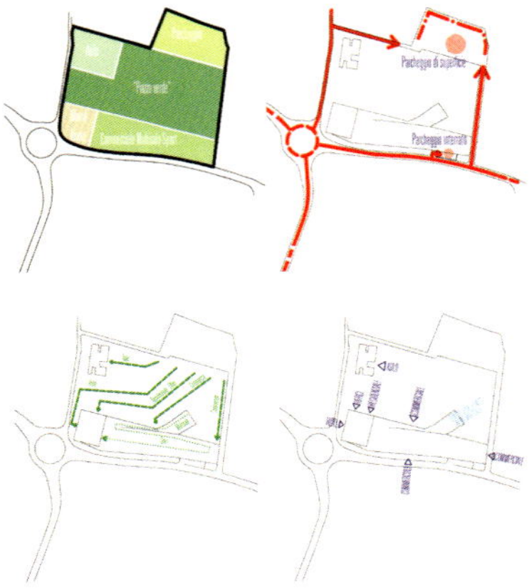

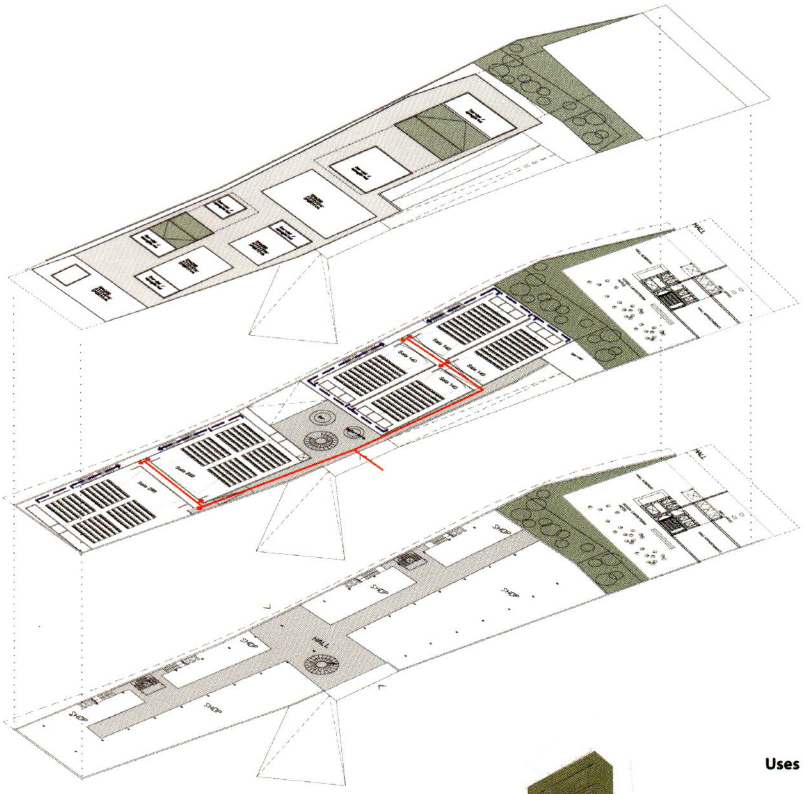

Facilities

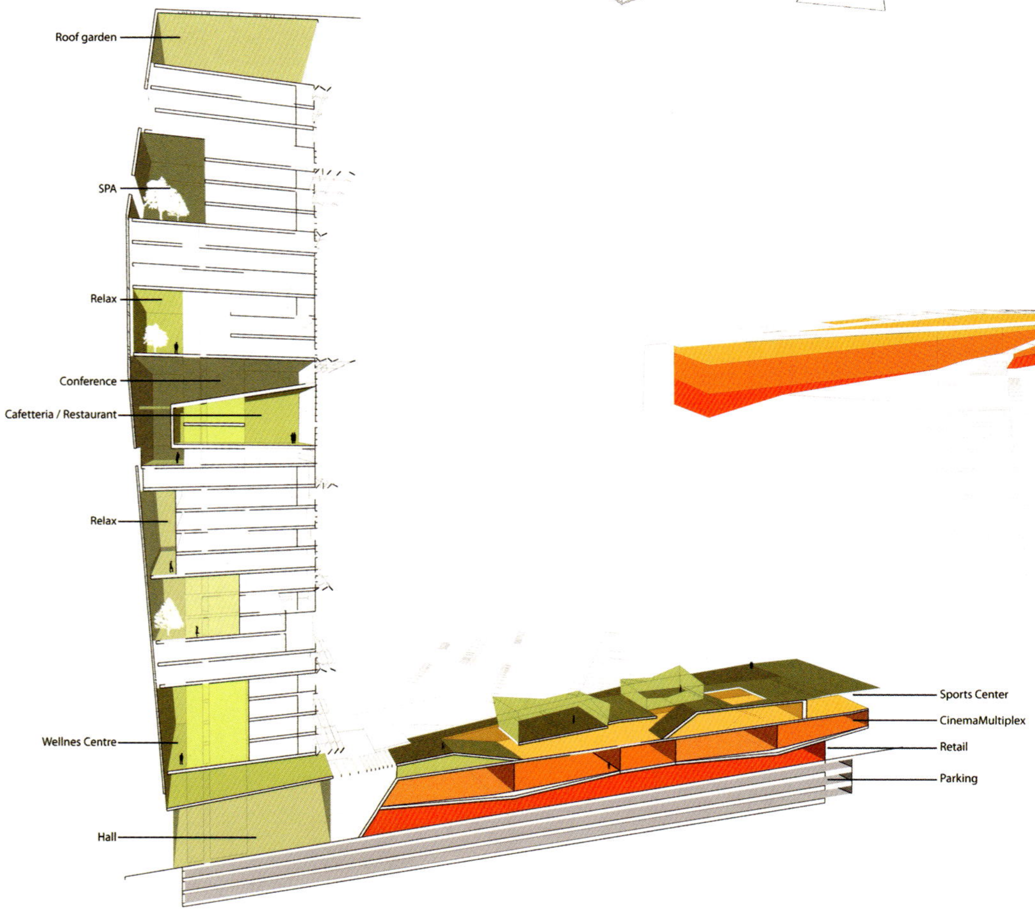

Uses

Vertical links

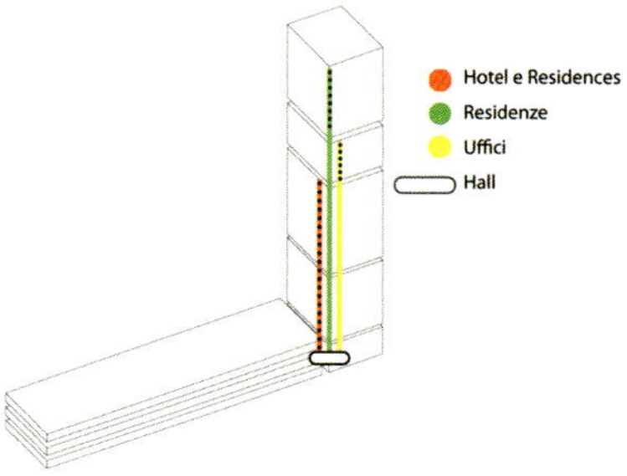

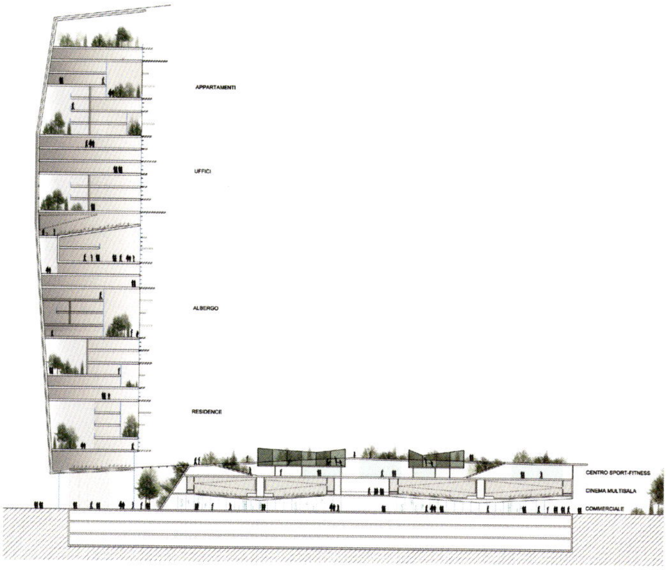

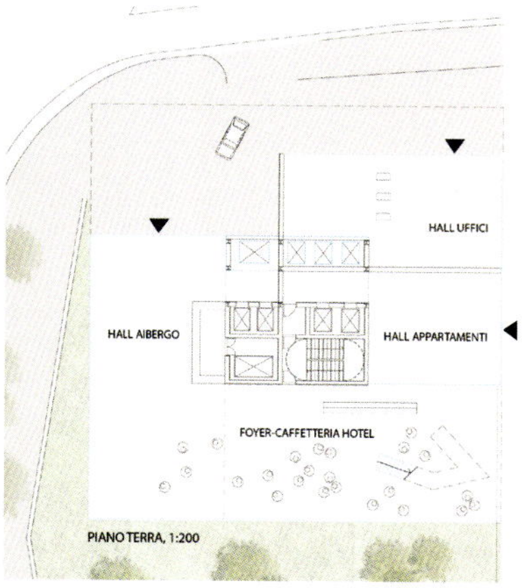

PIANO TERRA, 1:200

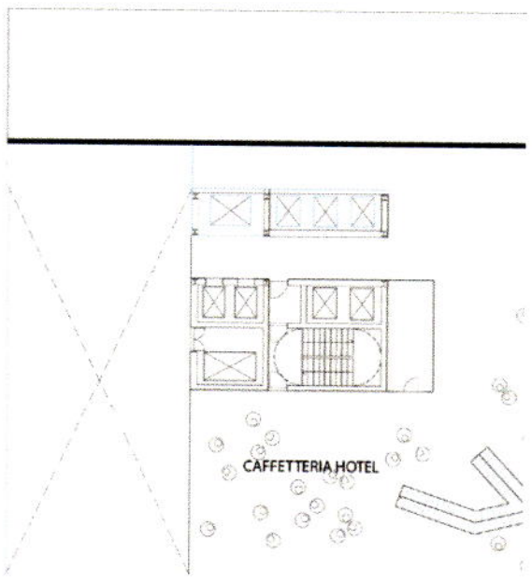

PRIMO PIANO, 1:200

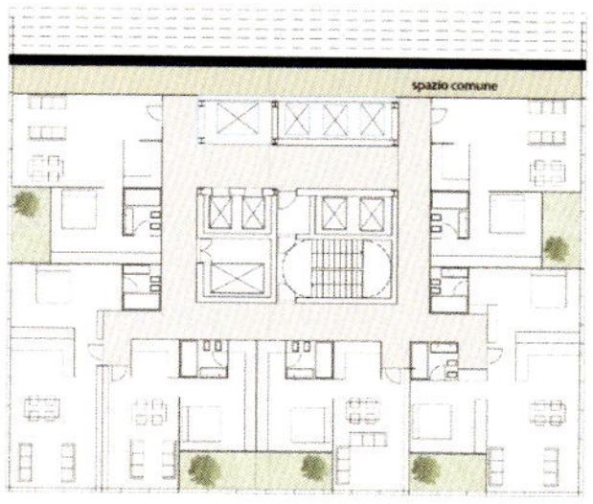

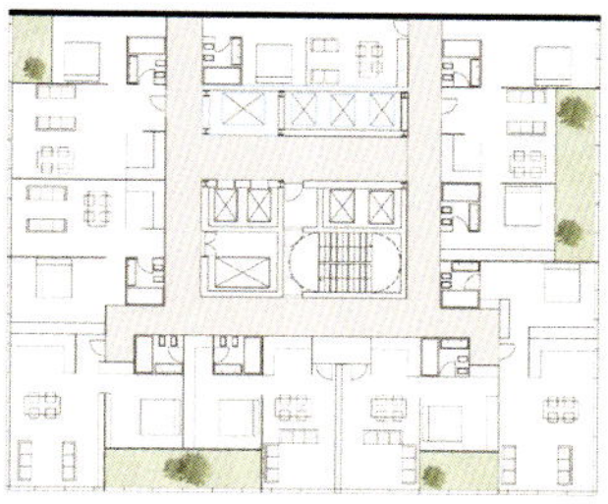

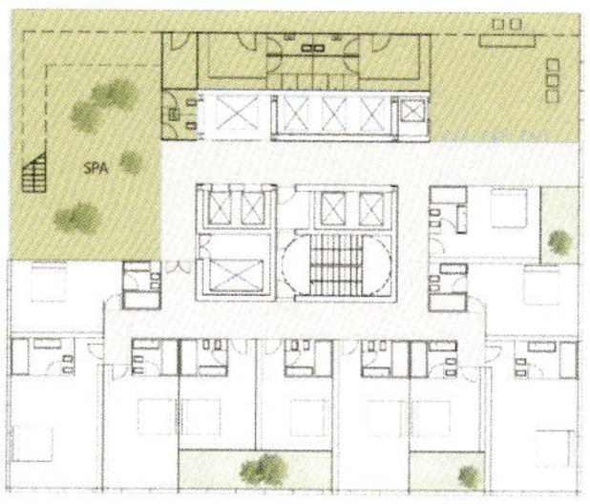

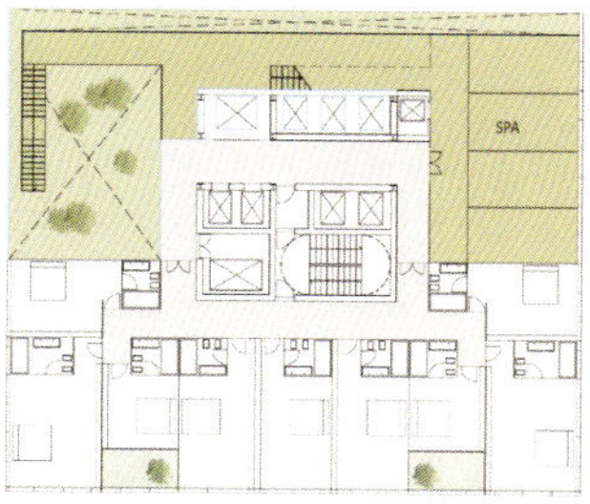

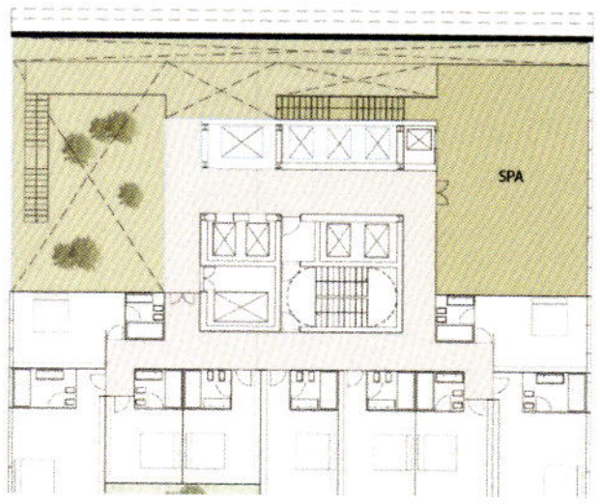

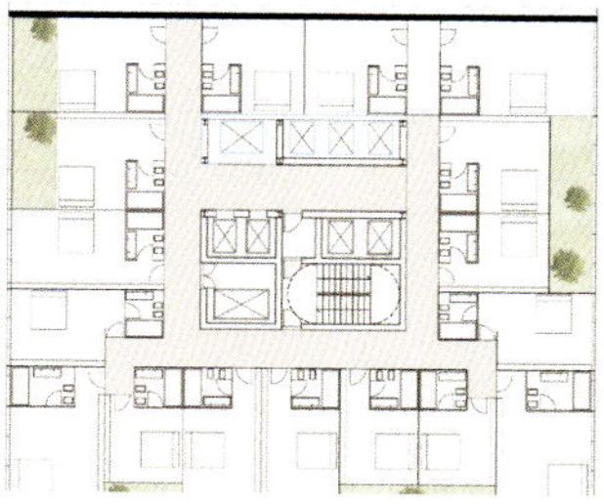

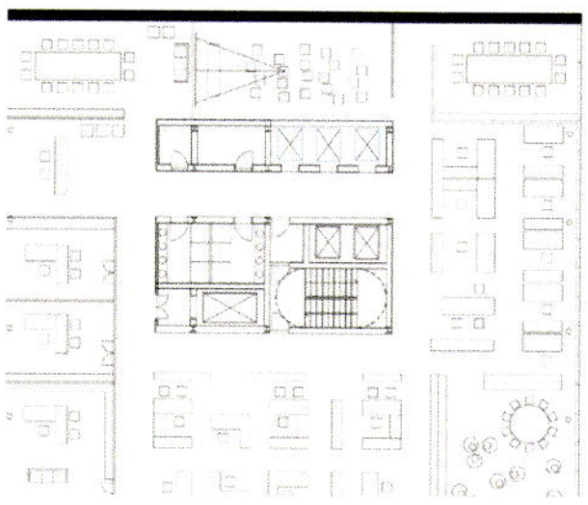

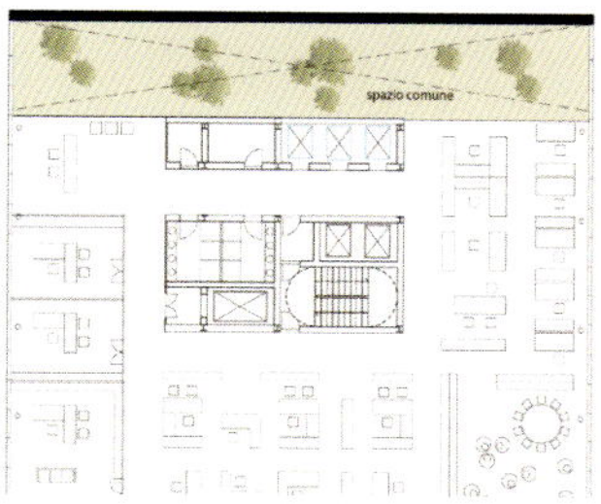

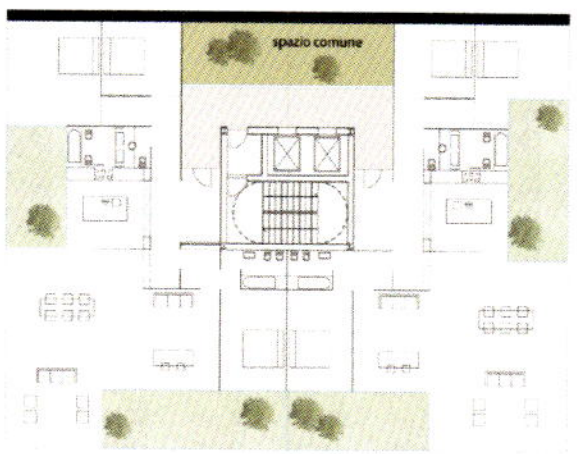

TIPO 1. 1:200

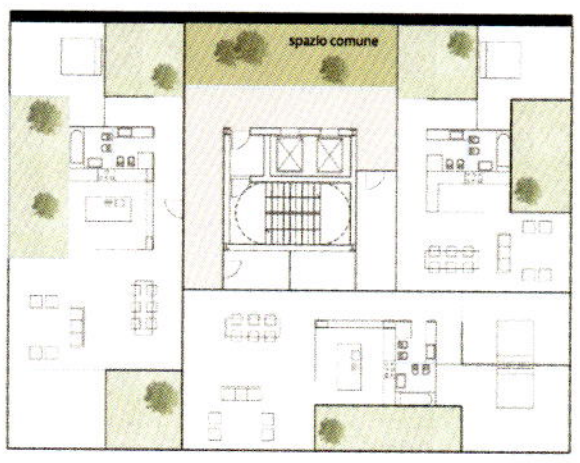

TIPO 2. 1:200

east gate

Taquin Housing Prototype

Taquin住宅模型

Invited Competition 'The House of the Future' Ance | Design: n!studio Susanna Ferrini, Antonello Stella | Team: Simone Bove, Gabriele Misso, Giulia Scaglietta, Alessandro Turina | Collaborators: Rocco Smaldone | Client: Ance – Associazione Nazionale Cosruttori Edili | Period: 2004
Model: Marco Galofaro

Is it possible to imagine the house like a form of time, as a system of variable configurations? Within contemporary research the house becomes a new instrument for the perception of reality, an ulterior gauge of the transformations of social and physical reality. The project was used to experiment with new processes of appropriating the space of dwelling by individuals. We have thus privileged the possibility of a continuous and intense process of 'appropriation' of the space of dwelling by its inhabitants, designing a mechanism that allows the inhabitant to continually vary the configuration of interior spaces, which slide along horizontal and vertical rails. Within the prototype dwelling the functional blocks can move in space, creating different internal spaces. The lines of movement of the functional blocks pre-ordain the trajectories of its inhabitants, 'following' their physical re-positioning in space as part of a network that defines successive loops and 'boxes' that house the various functions of dwelling. In addition to the idea of internal flexibility, the project is also aimed at suggesting the possibility of the co-existence, within a house seen as an industrial and prefabricated prototype, of different 'means' of dwelling: using and inhabiting the roof of the house as a new dimension of collective existence; vertically 'shifting' internal functional blocks in order to render the exterior space 'more private' and the ability to inhabit the finite dimension of the hyper-functional interior cells or, once again, to 'appropriate' and give meaning to the 'neutral' spaces located between the functional blocks.

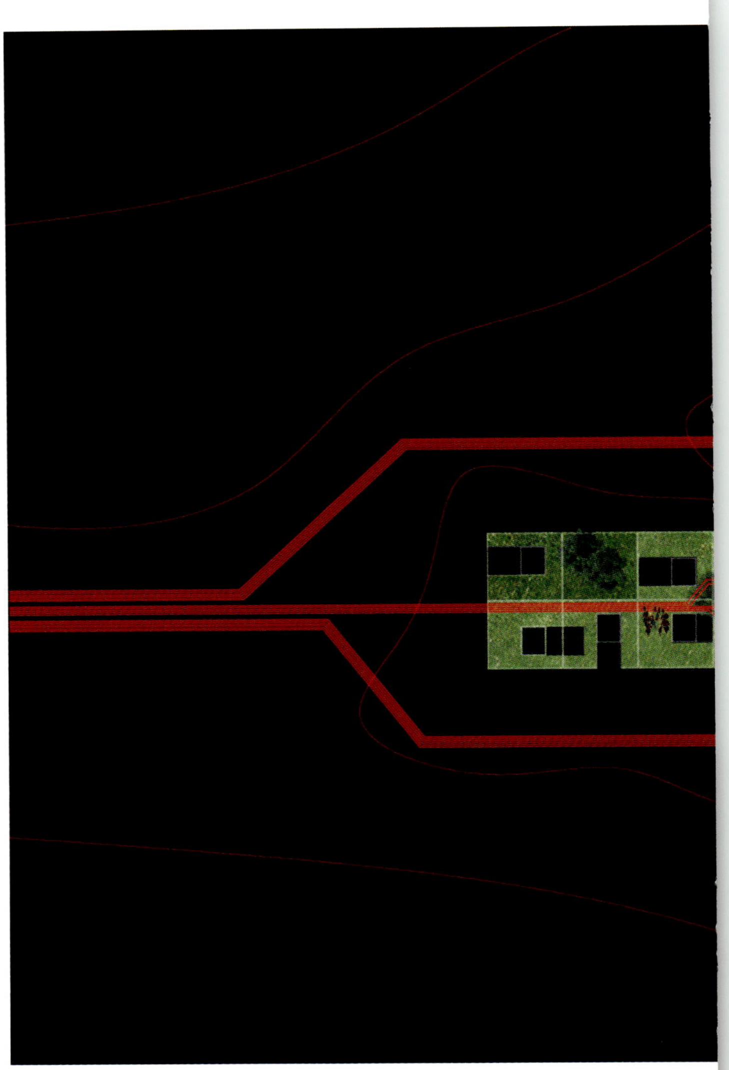

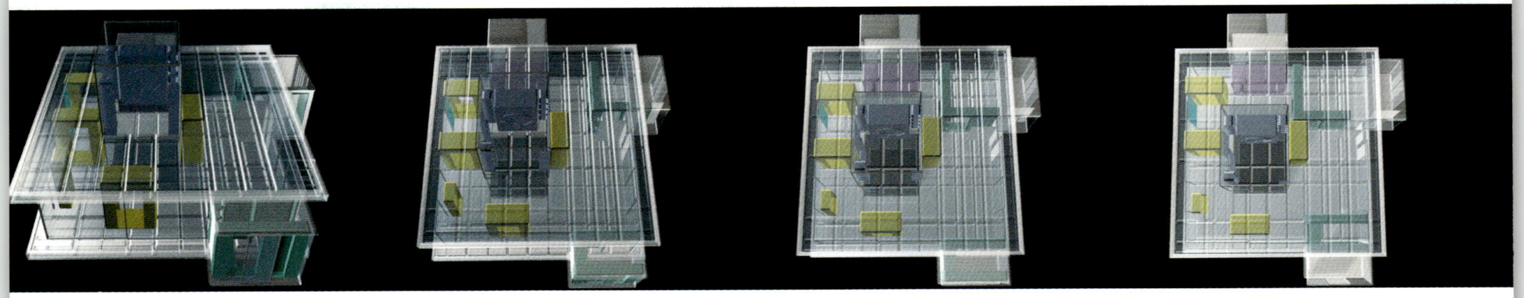

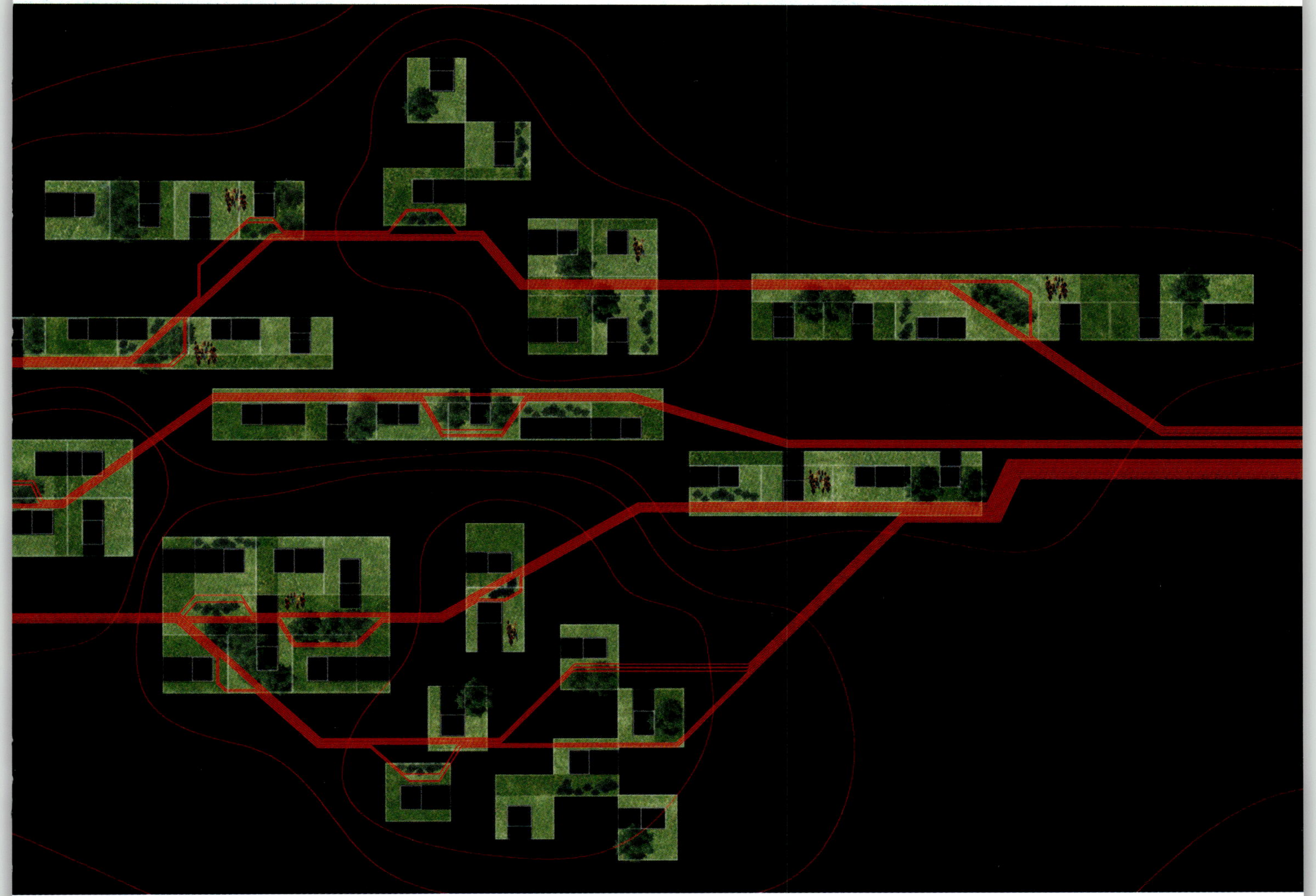

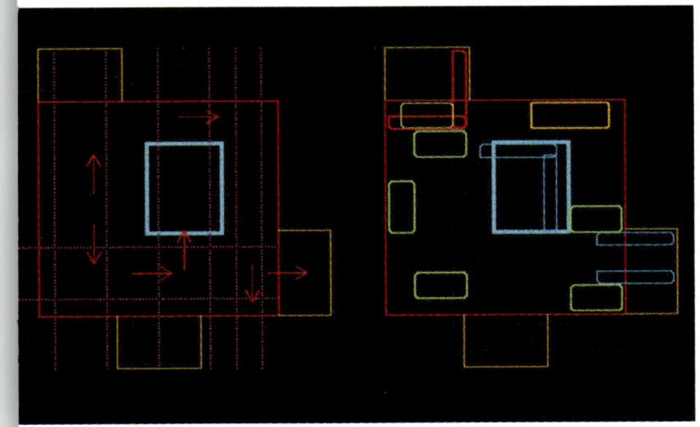

是否可能将住房想象成一种具有时间属性的形式，一种具有可变配置的系统?在当代研究中，住房成为了人们感知现实的一种新工具，一种衡量社会和物理现实的变化的隐含标尺。该项目使人们经历了合理分配个人居住空间的新流程。我们因此特别提供了使居民不断且大范围“分配”其居住空间的流程，设计出了一种可以使居民不断改变其内部居住空间配置的机制，这种空间模块可以在竖直及水平的轨道上滑动。在这种居住模型中，各种功能模块可在空间中移动，营造出不同的内部空间格局。这种功能模块的移动线路预先规定了其住民的移动轨迹，“随着”其物理空间的重新定位，形成了一种界定了连续的回路和“盒子”的网络，以容纳各种居住功能模块。除了内部可变性的概念之外，本项目还旨在为各种居住“手段”在作为工业和预制安装建筑模型的房间中的共存方式提供建议。利用及占据房屋的屋顶为这种共存提供了新的维度；竖直“移动”内部功能模块以使外部空间“更加隐私”并占据功能高度发达的内部空间的有限容积；或者再次“分配”那些位于功能模块之间的“中性”空间并赋予其新意义。

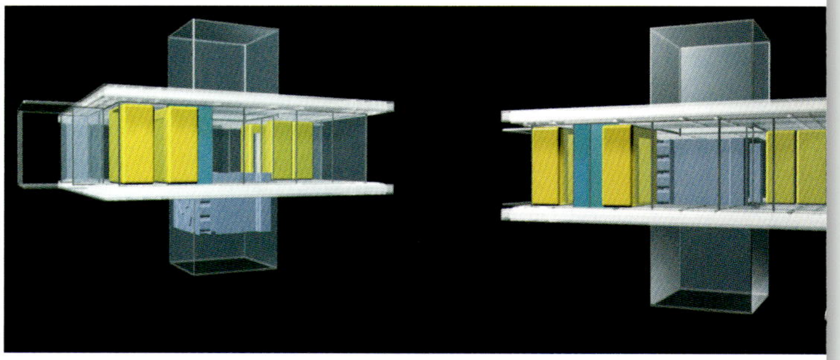

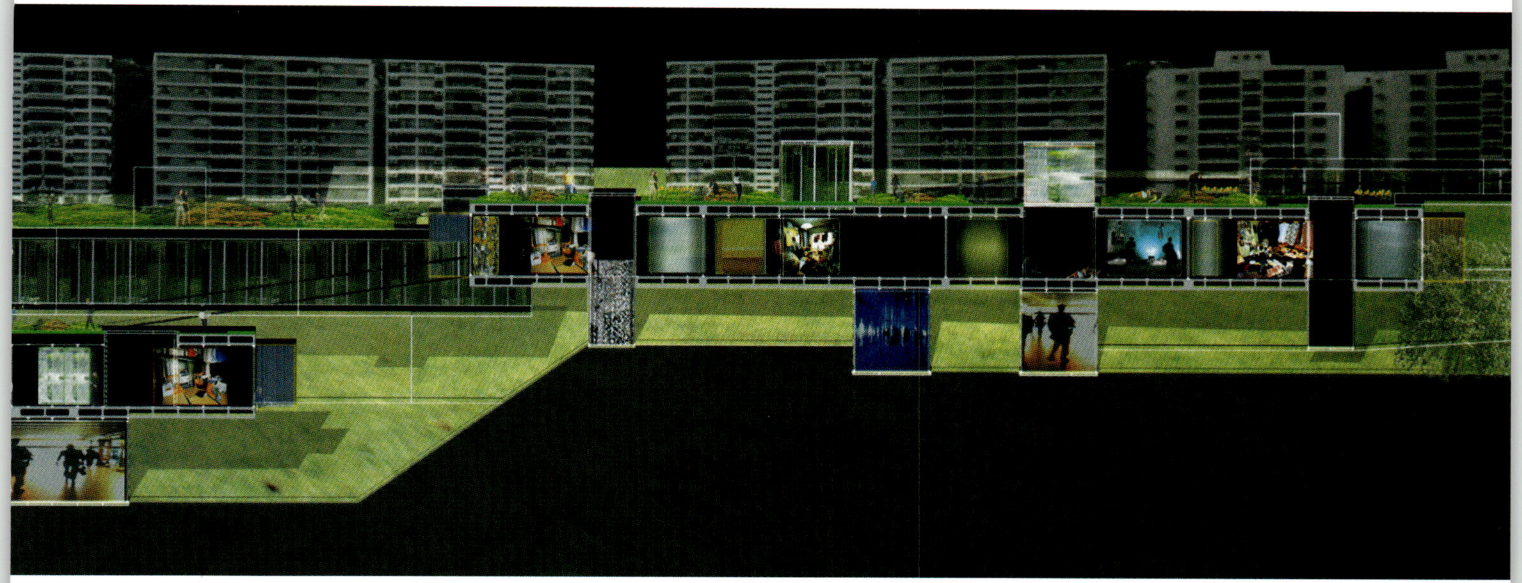

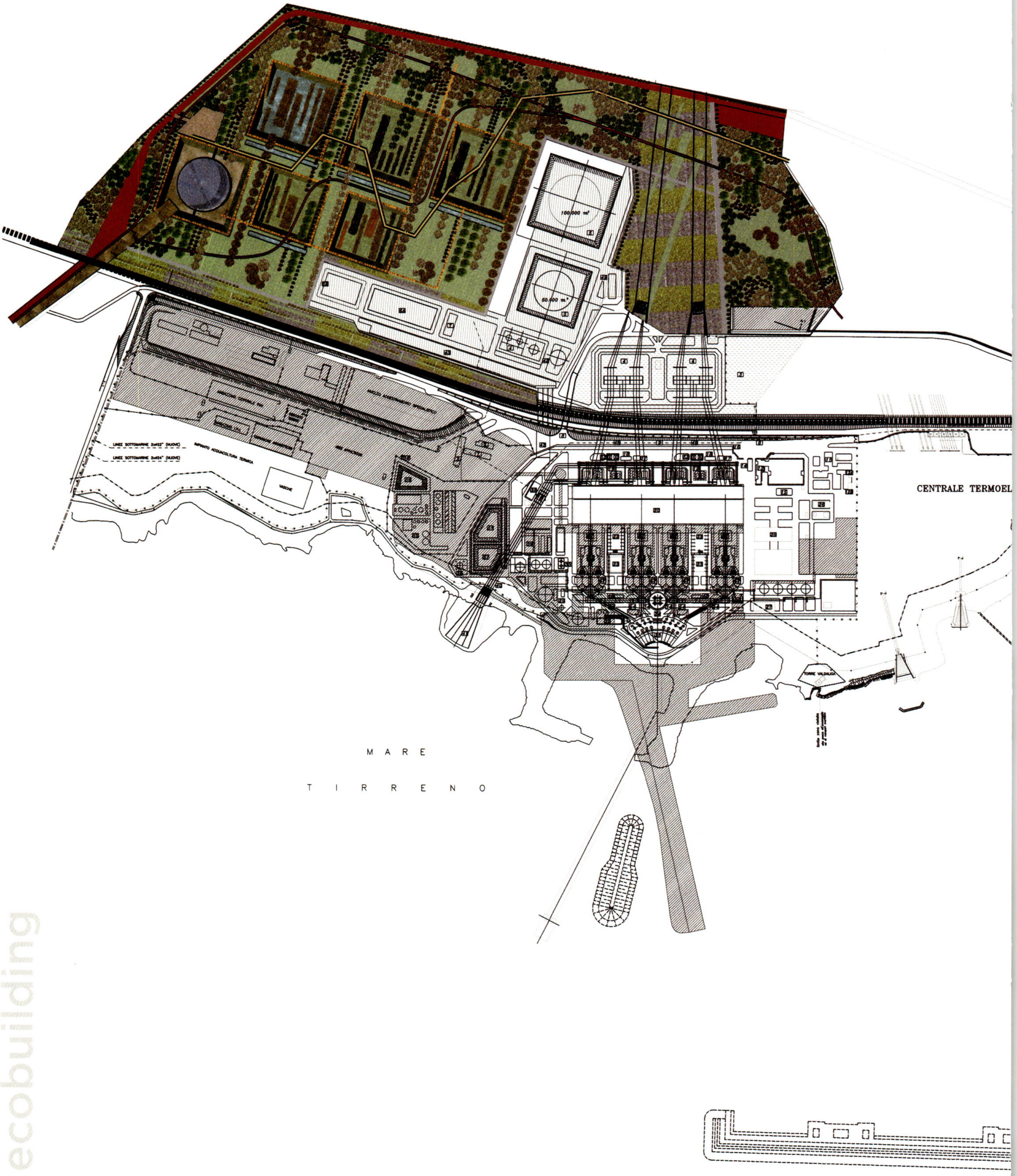
CENTRALE TERMOEL
MARE
TIRRENO

eservoir Park Civitavecchia

维塔韦基亚储油库公园

onmental Project for the Enel Power Station in Civitavecchia | Design: n!studio Susanna Ferrini, Antonello Stella | Consultants: Daniela
rini, Ippolito Pizzetti, Umberto Pessolano | Collaborators: Anna Cornaro, Valerio De Divitiis, Barbara Annunziata, Chiara Barale, Fabio Bonfa',
ara Falcone, Fabrizio Federico | Client: Enel Produzione S.p.a. | Photographer: Luigi Filetici | Period:2002

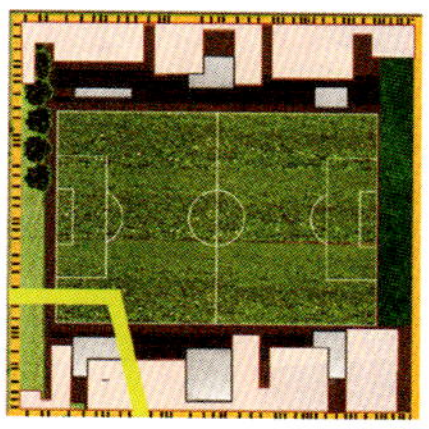

The project is an advanced feasibility study that is part of the plan for the conversion of the electrical power plant in civitavecchia, from combustible oil to coal. This delicate shift involves the decommissioning of the large oil reservoirs and the conversion of the area into a park. The project is focused on an idea of park that, given the proximity of the large infrastructure, even with all of its naturalistic connotations, maintains the memory of its original use, together with the new functions related to leisure activities. The first perplexity in satisfying the client's requests was related to the vocation of the site: if an inhabitant of civitavecchia wishes to enter into contact with nature, he can make use of such a vast and qualitative territory, from the coast inland towards the hills of tolfa, that he would certainly not be attracted by an area, even if transformed and environmentally requalified, that sits beside an electrical power plant. In light of this reflection the project focuses on a more 'utilitarian' use of the large reservoirs, imagining diverse functions of entertainment and attraction for leisure activities, from sports to 'environmental education'. The basic idea is that of significantly transforming the area, while maintaining its original signs: the large, square reservoir, 120 x 120 metres (larger than a soccer field), once used as the emergency containment area in the event of a leak of combustible oil from the tanks, will be maintained and used to host new functions, with the perimeter paths located atop the perimeter banks. The project also features various hypotheses for the reuse of one of the 9 tanks, with a diameter of 120 metres: from a greenhouse to the site of performances.

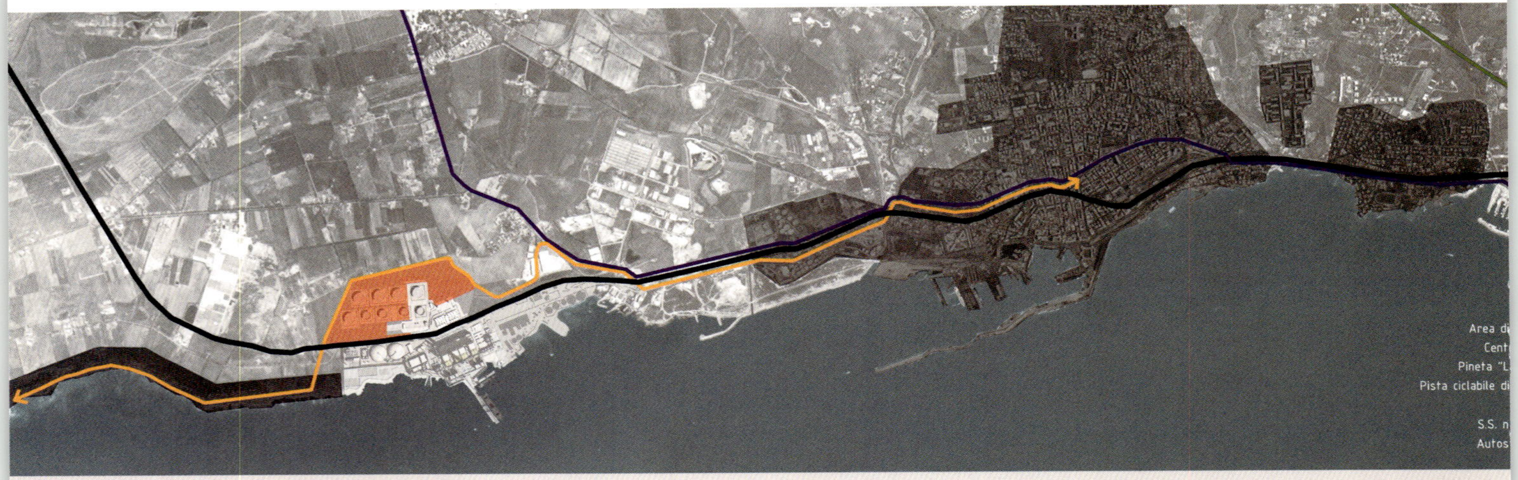

Inquadramento

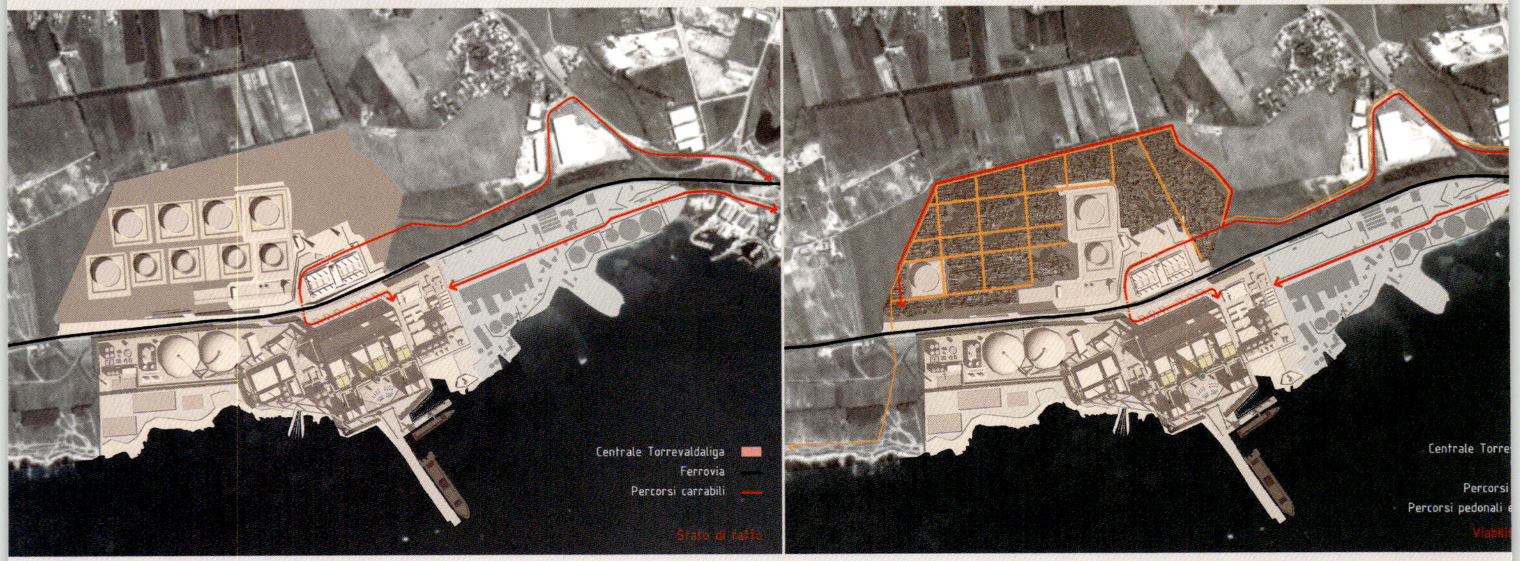

Schema dell

BONIFICA DELL'AREA DEI SERBATOI. IL PARCO DELLA CENTRALE

MANTENIMENTO E RIUSO DEL SERBATOIO ESISTENTE

DIVERSIFICAZIONE DEL TESSUTO "VERDE" DEL PARCO

GLI INVASI DEI SERBATOI RECUPERATI A SPAZI VERDI

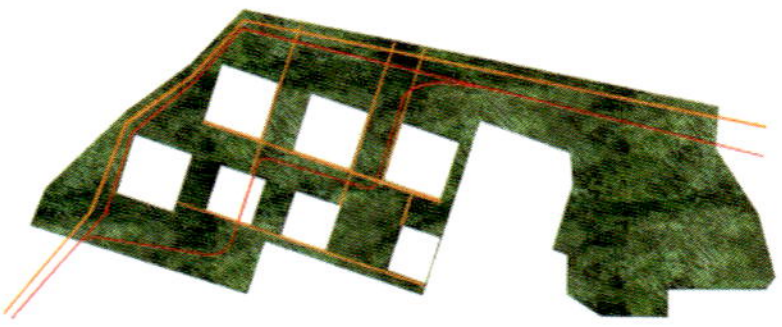

INGRESSI, PERCORRIBILITA' E PERCORSI DEL PARCO DELLA CENTRALE

TEMATIZZAZIONE DELLA SUPERFICIE DEL PARCO

RIUSO FUNZIONALE DEGLI INVASI DEI SERBATOI

LA DIVERSIFICQAZIONE DELLE ESSENZE ARBOREE:
ZONE FILTRO, FILARI ARBOREI, SUPERFICI A PRATO

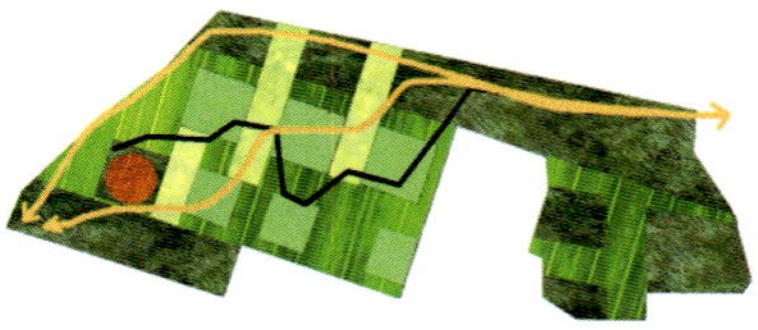

PROGETTO DEL PARCO DELLA CENTRALE

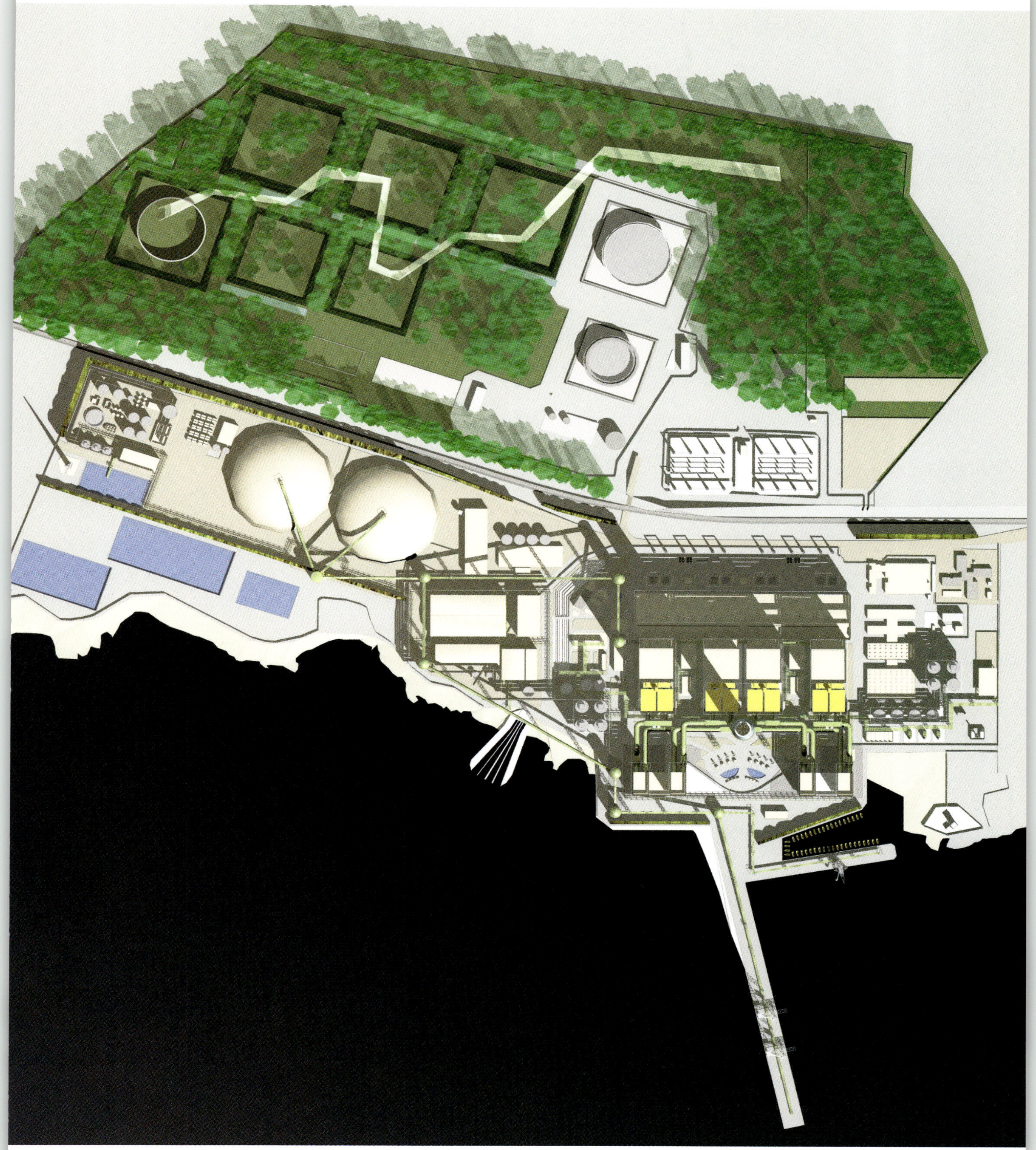

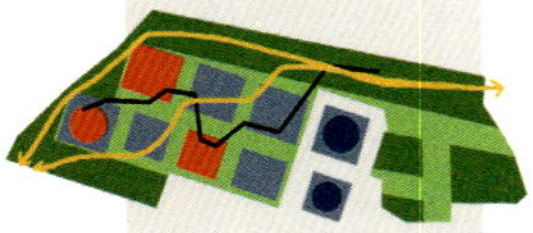

ELEMENTS OF THE PARK

VERDE

banda boscata

formazioni arbustive

canne ed erbe perenni

COLLEGAMENTI

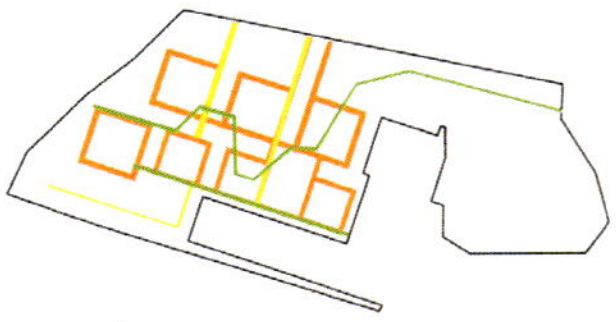

percorsi

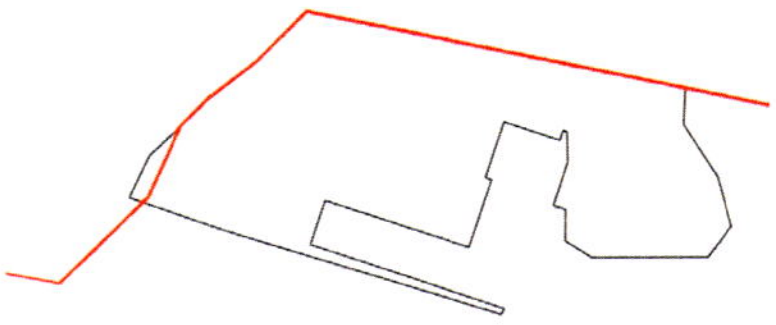

pista ciclabile

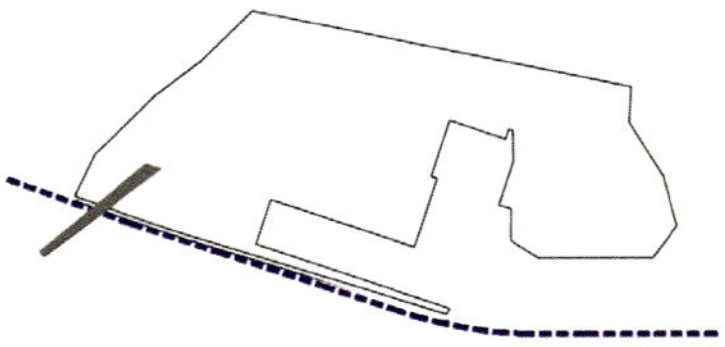

collegamento con pineta

FUNZIONI

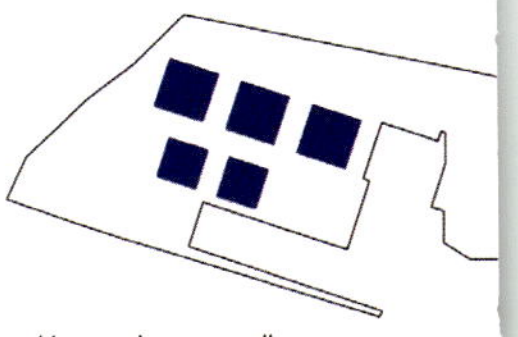

attrezzature per il parco

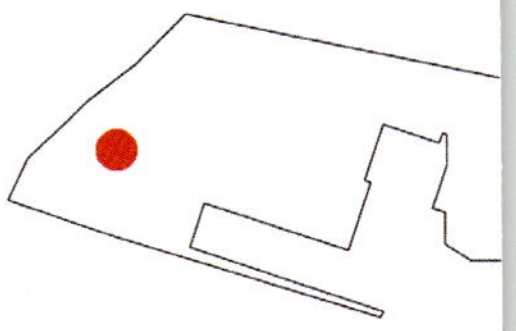

serbatoio riutulizzato

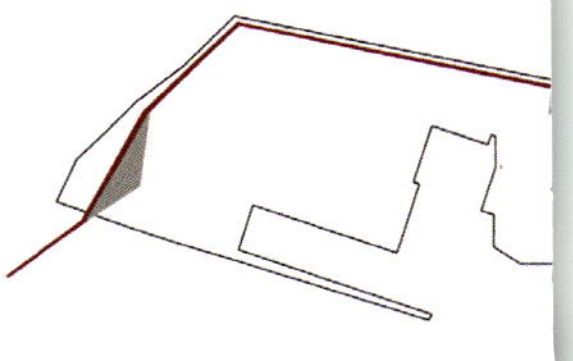

parcheggi e viabilità

REUSE OF THE TANKS: POTENTIAL CONFIGURATIONS

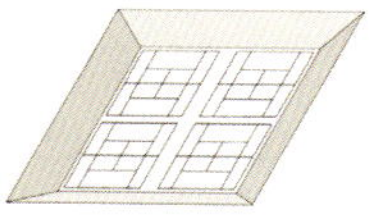

attrezzature sportive

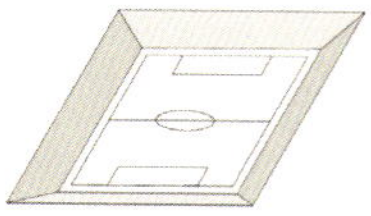

campo da calcio

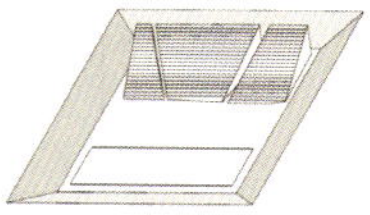

teatro all'aperto

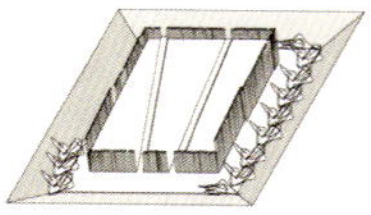

ecosistema d'acqua dolce

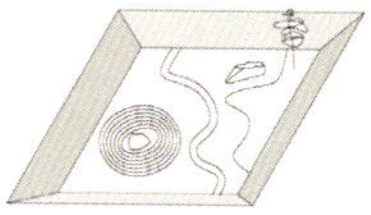

giardino tematico

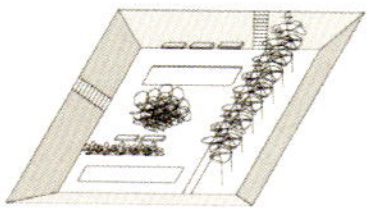

piazza attrezzata

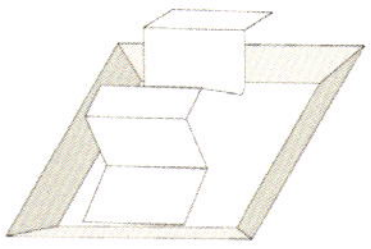

attrezzature tempo libero

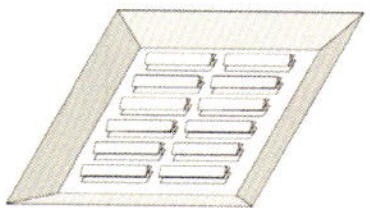

allestimenti temporanei

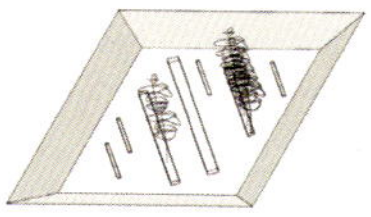

giardino didattico

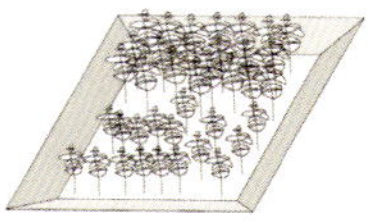

parco

PISTA DI SKATEBOARD
PISTA CICLABILE
CINEMA ALL'APERTO

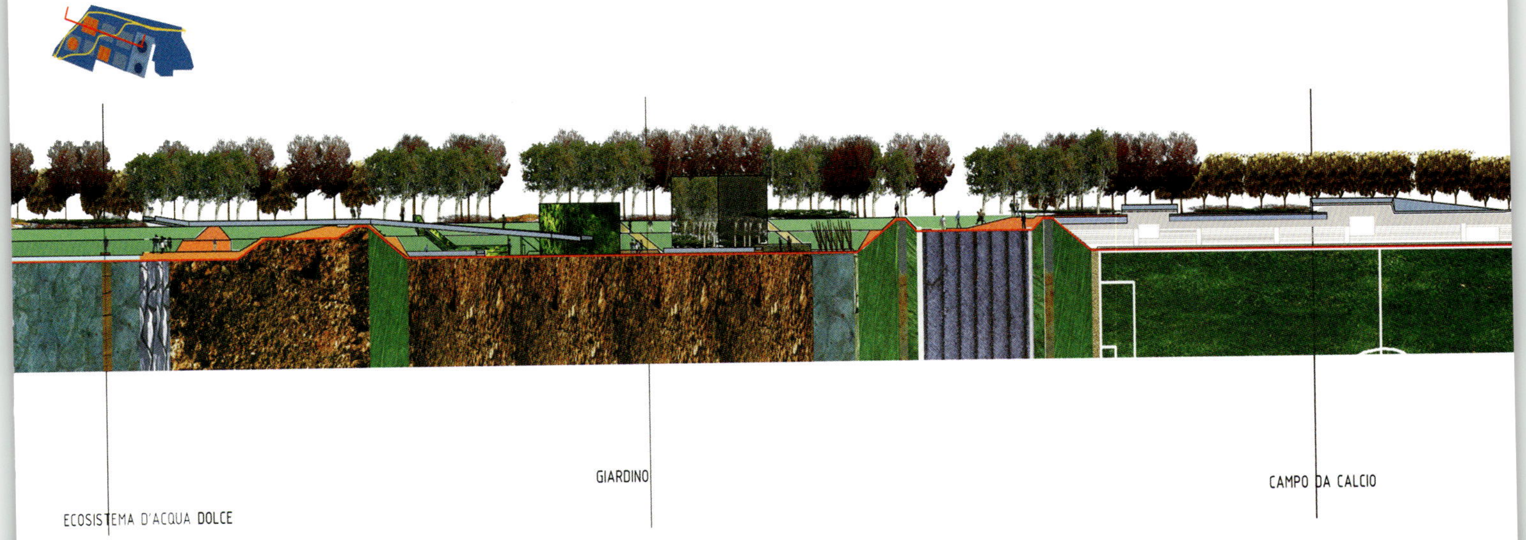
GIARDINO
CAMPO DA CALCIO
ECOSISTEMA D'ACQUA DOLCE

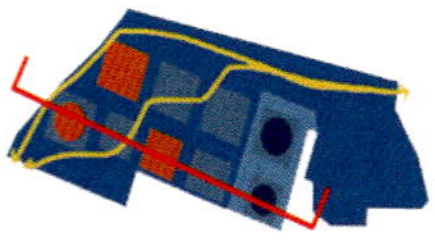

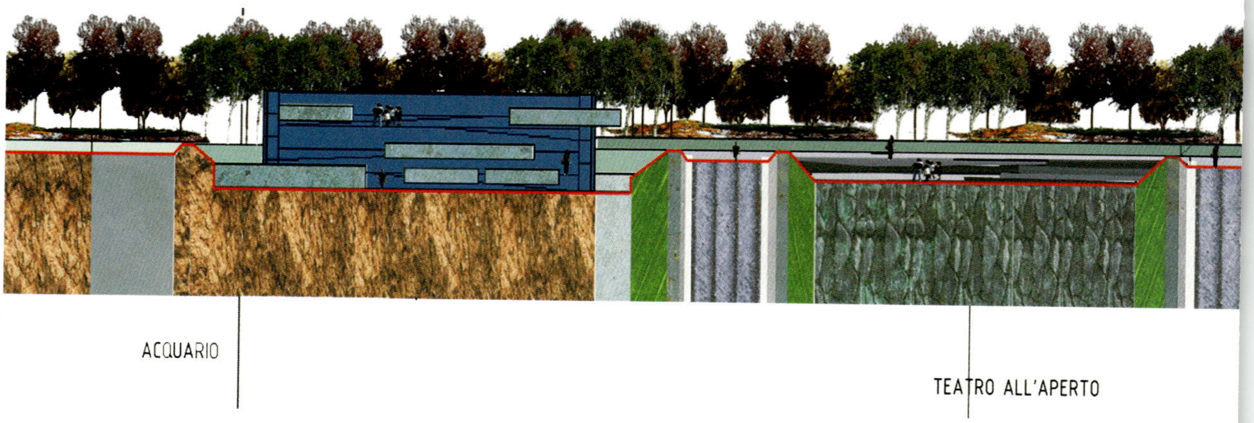
ACQUARIO
TEATRO ALL'APERTO

tennis
nuoto
pallavolo
cinema
teatro
espos

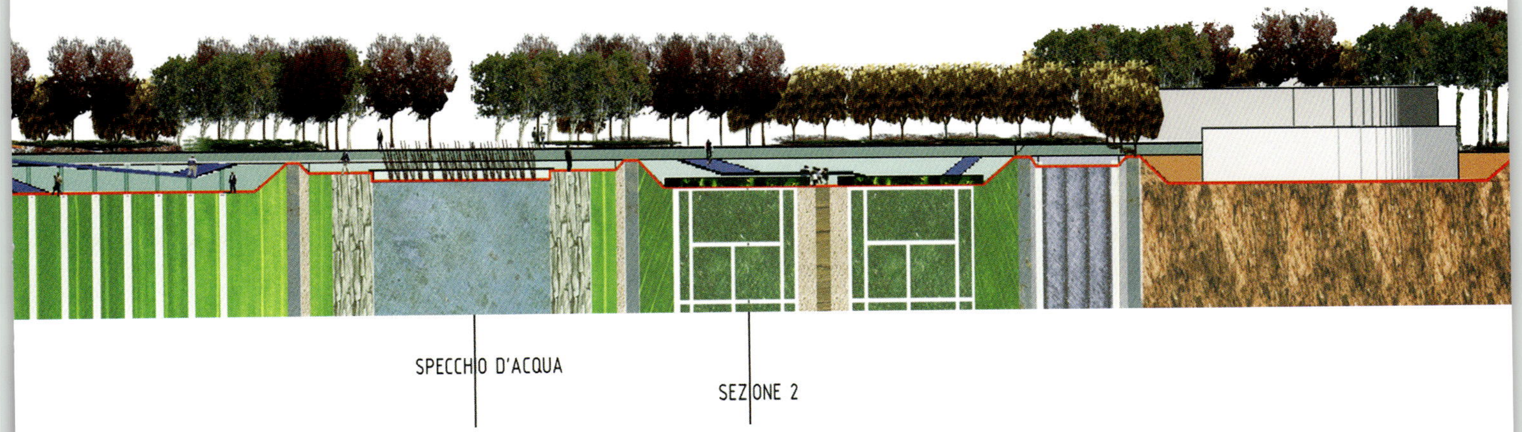
SPECCHIO D'ACQUA
SEZIONE 2

orti
botanica
palestra di roccia
acquario
serra
parcheggi

STAGNO ARTIFICIALE NATURALIZZATO CON INSERIMENTO
DI TAXA ANIMALI E VEGETALI IDONEE.

本项目是一项高级可行性研究，它是将奇维塔韦基亚发电站从燃油发电转为燃煤发电规划项目的一部分。此项极难处理的改造项目涉及停止运行其大型储油库并将该区域改造成为一个公园。此项目重点集中在对此公园进行设计，使其在临近大型基础设施的情况下，即便使该区域具有了所有自然主义内涵，仍能够使人们保留对其最初用途的回忆，同时也要增添与休闲活动相关的新功能。在满足客户的要求时所遇到的第一个难题与该场地的职业用途有关：如果奇维塔韦基亚的居民想要步入自然，他们尽可以享受从沿海内陆一直延伸到Tolfa山的广袤地域，而位于一座发电厂旁边的地区对他当然没有什么吸引力，即便该区域已经得到改造，环境质量得到了提高。考虑到这一点，该项目更多地集中到了大型储油库的“功用”方面，描述了从运动到“环境教育”的各种各样娱乐和休闲活动功能。其基本想法就是在大力改造该区域的同时保留其最初标记：这种曾经在燃油罐泄漏时被用作紧急储油区、大小为120米 x 120米的大型方形储油库（比一个足球场还要大），将被用于开发新的功能。这种储油库的周边堤坝上有着环绕其周边的道路。此项目还包括为重新利用9个直径为120米的油罐空间提出各种设想：从温室到表演场地。

E
ATTREZZATURE LUDICHE
RECUPERO DEGLI INVASI CON SPECIE ARBOREE
PERCORSO PEDONALE IN QUOTA

ecobuilding

Amalfi Tourist Centre

Amalfi旅游中心

Waterpower Renewal Strategy for the Mulini Valley in Amalfi and Scala Recovery of the Former Soap Factory in Amalfi | Design: n!studio Anna Ferrini, Antonello Stella Architetti Associati | Team: Simone Bove, Giulia Scaglietta, Gabriele Misso, Davide Gamba | Coordinator: Luigi Centola, Centola & Associati | Client: University of Salerno | Period: 2004 | Award: Holcim European Awards 2005

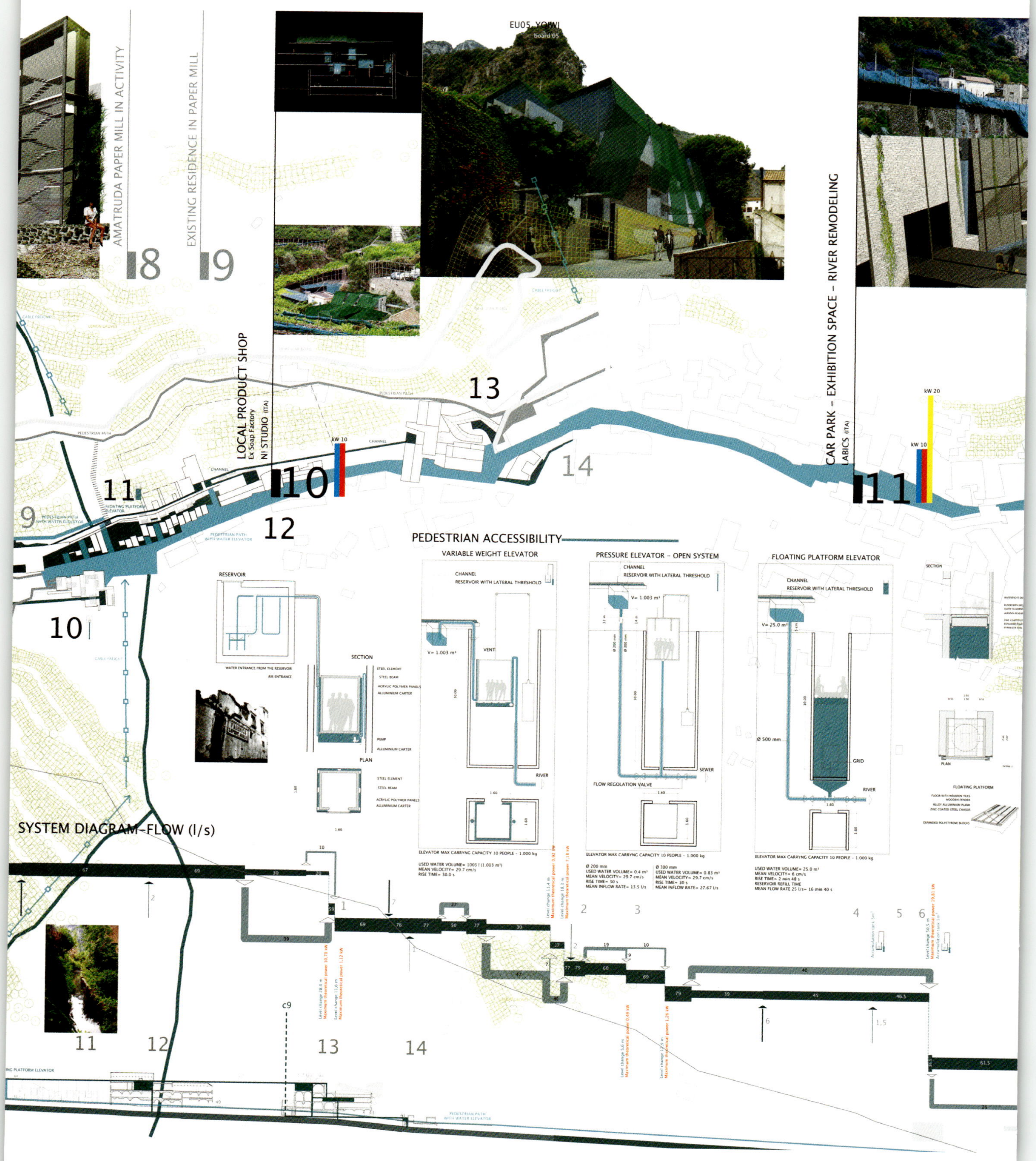
EU05_YOIWI
board 05
AMATRUDA PAPER MILL IN ACTIVITY
EXISTING RESIDENCE IN PAPER MILL
8
9
LOCAL PRODUCT SHOP
Ex Soap Factory
N! STUDIO (ITA)
CAR PARK – EXHIBITION SPACE – RIVER REMODELING
LABICS (ITA)
kW 10
kW 20
10
11
12
13
14
PEDESTRIAN PATH
CHANNEL
CABLE FREIGHT
PEDESTRIAN PATH WITH WATER ELEVATOR
FLOATING PLATFORM ELEVATOR
PEDESTRIAN ACCESSIBILITY
VARIABLE WEIGHT ELEVATOR
PRESSURE ELEVATOR – OPEN SYSTEM
FLOATING PLATFORM ELEVATOR
RESERVOIR
SECTION
PLAN
CHANNEL
RESERVOIR WITH LATERAL THRESHOLD
V= 1.003 m³
V= 25.0 m³
VENT
RIVER
SEWER
FLOW REGOLATION VALVE
GRID
Ø 500 mm
ELEVATOR MAX CARRYNG CAPACITY 10 PEOPLE - 1.000 kg
USED WATER VOLUME= 1003 l (1.003 m³)
MEAN VELOCITY= 29.7 cm/s
RISE TIME= 30.0 s
Ø 200 mm
USED WATER VOLUME= 0.4 m³
MEAN VELOCITY= 29.7 cm/s
RISE TIME= 30 s
MEAN INFLOW RATE= 13.5 l/s
Ø 300 mm
USED WATER VOLUME= 0.83 m³
MEAN VELOCITY= 29.7 cm/s
RISE TIME= 30 s
MEAN INFLOW RATE= 27.67 l/s
USED WATER VOLUME= 25.0 m³
MEAN VELOCITY= 6 cm/s
RISE TIME= 2 min 48 s
RESERVOIR REFILL TIME
MEAN FLOW RATE 25 l/s= 16 min 40 s
FLOATING PLATFORM
SYSTEM DIAGRAM–FLOW (l/s)
c9
11
12
13
14

EU05_YQIWJ
board 06
AUDIOVISUAL RESEARCH CENTER
Ex Pansa Paper Mill
MARANO (ITA)
kW 10
12
CHATEAUX RELAIS – POOL AND LEMON TERRACES
Ex De Luca Paper Mill
MIRALLES TAGLIABUE – EMBT (ESP)
kW 10
13
AMALFI PAPER MUSEUM
14
CATHEDRAL
PIAZZA
FOUNTAIN
AMALFI
9 10 11 12 13 14
SEA
Holcim Awards for Sustainable Construction – European Winner
www.waterpoweramalfi.com
WATERPOWER
Renewal Strategy for the Mulini Valley
Amalfi & Scala
Main Authors
LUIGI CENTOLA – CENTOLA & ASSOCIATI
MARIAGIOVANNA RIITANO – UNIVERSITÀ DI SALERNO
Client
PROVINCIA DI SALERNO
SITE PLAN
100m 50 20 10 0
N
Provincia di Salerno
Università degli Studi di Salerno

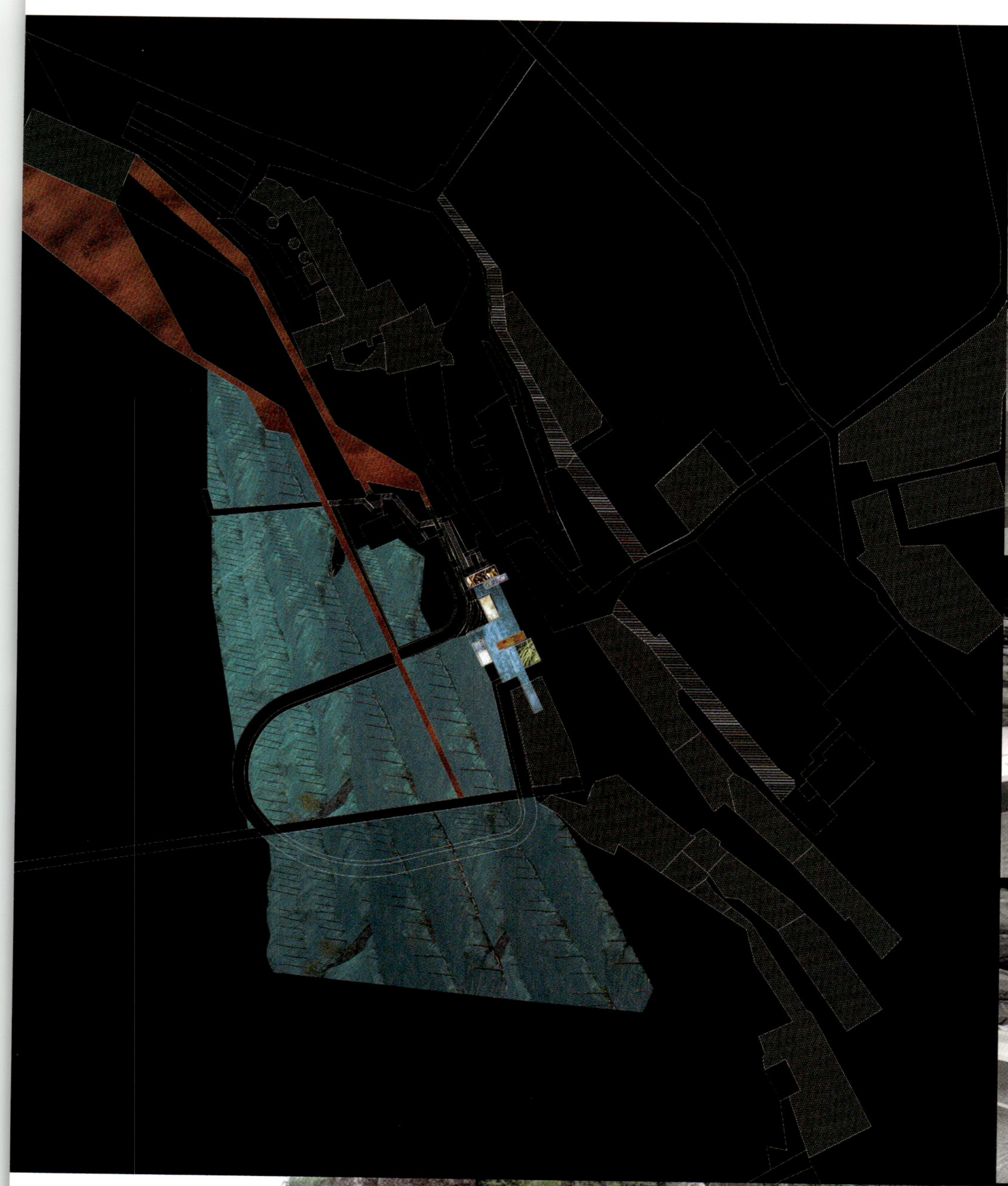

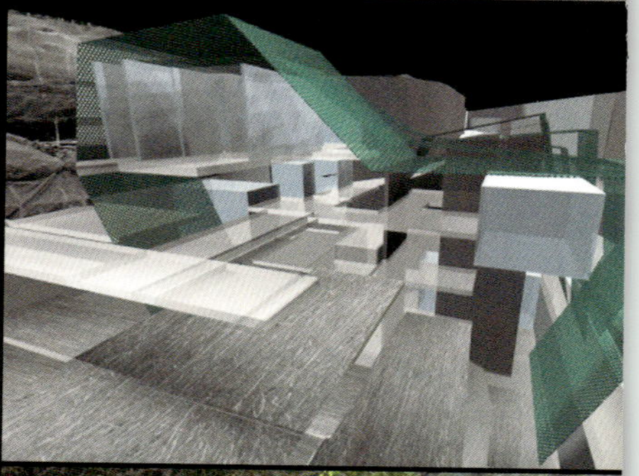

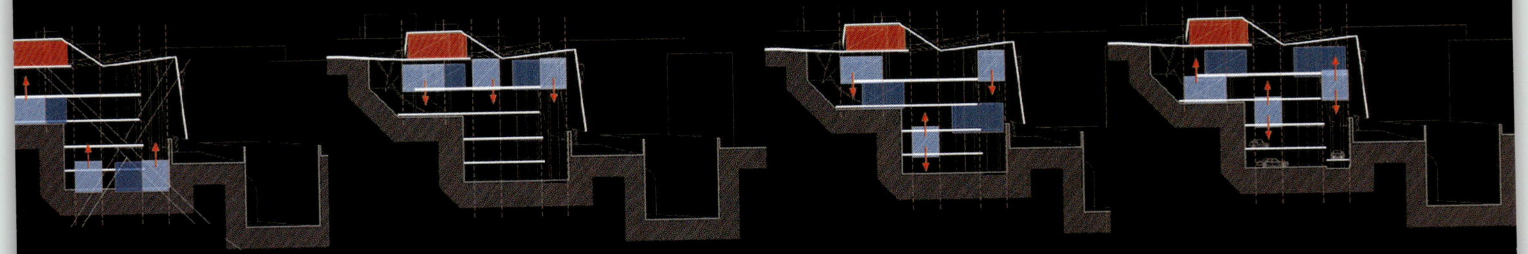

Within the complex longitudinal system of the 'valle dei mulini', the former amalfi soap factory represents a missing piece of the puzzle in the succession of the important remnants of industrial archaeology found in the area, currently in need of recovery. The structure of the project is the result of the configuration of the landscape itself, where stone retaining walls, terraces and tarpaulins protecting lemon groves create a fascinating image, capable of representing the layering of the culture of this particular area of the mediterranean. The design is thus based on an type of architecture that privileges 'subtraction' over the re-proposal of an imposing volume, creating an architecture of the ground plane, composed of articulated planes on various levels that are 'protected' – like the terraced lemon groves – by a series of folded roofs, similar to large sails or tarpaulins. As part of this approach, the intervention is organised according to systems, which tend to integrate themselves within the dynamic structure of a machine for exhibitions, with a variable configuration. The first system is composed of the first two levels, occupied by a mechanised parking structure. The second is composed of display cases for local products, which move vertically together with the movement of the elevators. The third system is that of the exhibition spaces, planes of modular and prefabricated structures. The variation of form, the reversible nature of the intervention, the articulation of the voids related to circulation and movement, the folds of the roof that capture the light and shadow of the landscape thus become the themes that contribute to the overall 'meaning' of the project.

在Valle dei Mulini复合型纵深系统中，前Amalfi肥皂厂是这一区域重要的老工业建筑遗迹中不可或缺的一部分，此区域目前正需要进行重建以恢复其原貌。此项目的结构是通过对该地区地貌本身进行布局配置而得到的，其用于保护柠檬林的石头护壁、阶地和油布形成了一道迷人的风景，完全体现出了这一独特地中海区域文化的层次感。于是，其设计便采用了一种注重“删减”而不是重新增加建筑规模的形式，创造出一种分为各种层次而又由相互连接的平面组成的接地层，各层均得到了一系列像大型船帆或篷布一样的折叠式屋顶的“保护”——就像阶地式的柠檬林一样。作为本方法的一部分，根据系统组织插入了各式结构，而这些插入结构都以不同的配置形式融入到了一座机器的动力结构中以完成展示目的。第一个系统由前两层组成，是机械化的停车结构。第二个系统由当地产品的展示柜组成，在竖直方向随电梯的走势而移动。第三个系统由展区、模块化预制安装结构的平面机构组成。其形式的变化、插入式结构的可逆转性、与流通和移动相关的各间隙之间的连接方式以及保障充分采光和遮阳的折叠式屋顶，成为了能够更加突出此项目整体“意义”的主题。

Green Parking Building

绿色停车场

Eco-façade for an Automated Parking Structure, Palermo | Design: n!studio Susanna Ferrini, Antonello Stella | Team: Maria Angela Corsi, Stefano Battaglia, Simone Bove, Paolo Rossi | Engineering: 3ti Progetti Italia | Client: Palermo Municipality | Period: 2007

ecobuilding

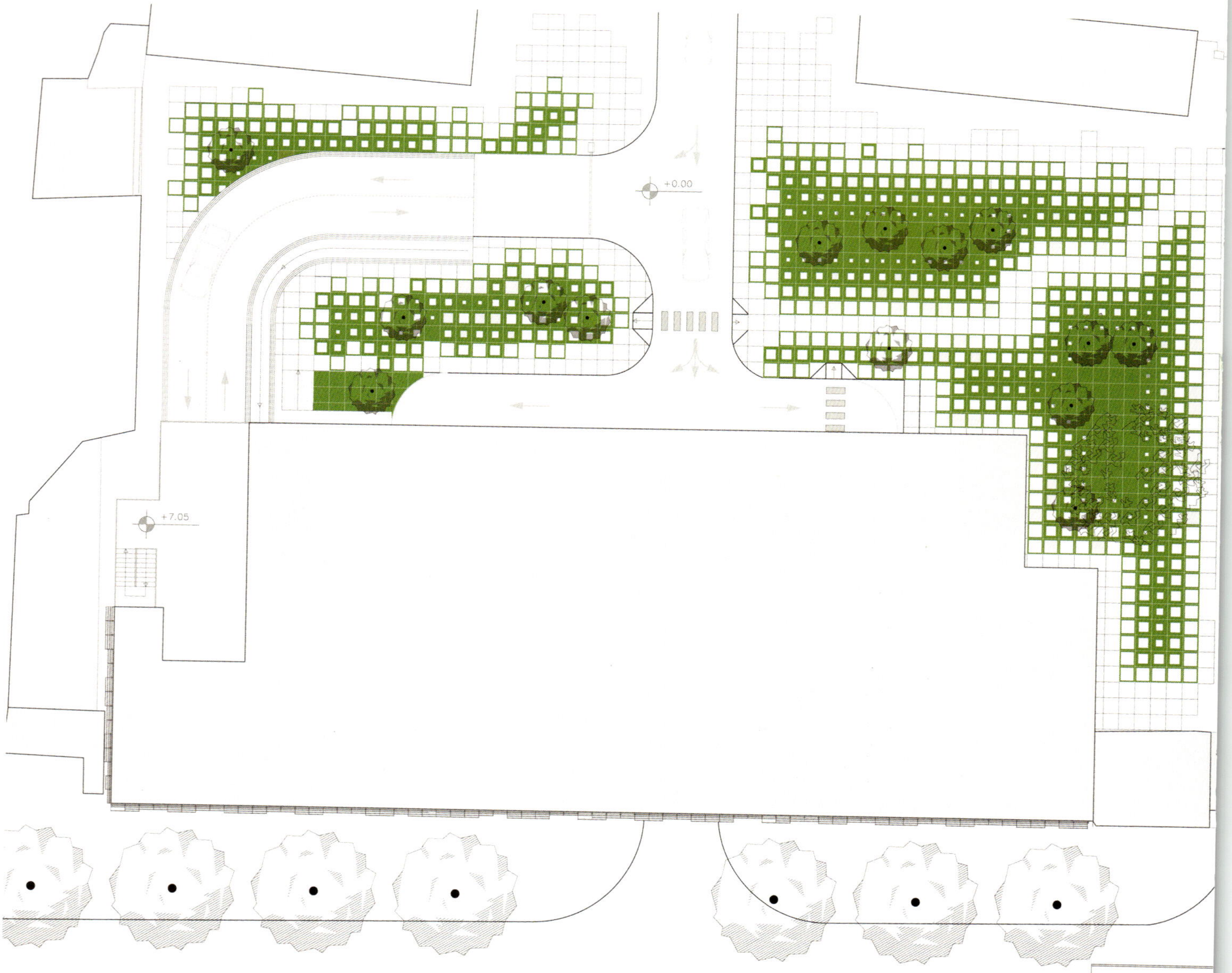
+0.00
+7.05

More than simplify fulfilling its role as a parking garage, the new building is an important architectural intervention that, while setting itself apart from the general appearance of this part of the city, is capable of becoming a highly characteristic element, precisely as a result of the insertion of its new façade. While it seeks a connection with its neighbouring buildings, the project is not an imitation of their architecture, but rather, thanks to its new façade, the transmitter of a highly contemporary image of the entire public square onto which it faces. The new building thus becomes an important element for re-design of the square. Along the Viale di Piazza Giulio Cesare, the rows of planted trees become an important environmental element, on par with the history of the buildings located here. As a result of their number and density, their visual impact will unavoidably become integrated with the new construction, hiding and revealing its façade with the changing of the seasons. The texture of the building is inspired by the profile of the tree foliage and ideally continued in the pattern designed by the overlapping exterior cladding panels that, as they project in three different orders, create a play of overlapping surfaces of varying densities, resulting in the intersection of the green treetops and the gridded skin. This mixture of natural and artificial elements creates an overall effect of lightness and progressive dematerialisation. Following an in-depth investigation of the project and the site, we chose to construct the pattern of the façade using a grid of fibreglass panels. This material, in addition to its aesthetic qualities of translucence and lightness, also offers an unlimited duration and requires no maintenance. Furthermore, its resistance to chemical and atmospheric agents makes it ideal for this part of the city, with its dense traffic and aggressive marine environment.

除了作为停车场之外，此新建筑还是一项重要的建筑设施，它同该城市这一区域的整体外观形象截然不同，正是其新颖的外表面结构设计使其成为了一座具有标志性特色的建筑。在同其周围建筑建立起联系的同时，本项目通过其新的外表面结构体现了其所面对的整个公共广场的非常现代的形象。因此新建筑成为了对此广场进行重新设计过程中的一个重要环节。沿着Viale di Piazza Giulio Cesare，一行行栽植的大树成为了重要的环保设施，与此处的历史性建筑交相辉映。因为树木多且密，其所造成的视觉效果将毫无疑问地融入到此新建筑中，随着季节的变化使该建筑的外表面变得若隐若现。此建筑物结构的灵感来自这些树木枝叶的轮廓，理想地按照其相互交叠的外部覆层板设计模式延伸，这些外部覆层板按三种不同的方式凸出来，形成了密度不一且相互交叠的表层，使绿色的树帽同其网格式表层交融在一起。这种自然和人工元素相互混合的形式造成了其整体的轻盈和逐步非物质化的效果。在对此项目及其现场进行了深入的勘察研究之后，我们选择采用玻璃纤维板栅格的形式建造其外表面结构。这种材料除了具有半透明和轻盈的美学质量之外，还经久耐用，无需维护。此外，它对化学品和大气成分具有很高的抵抗力，是能够适应该城市繁重的交通环境和海洋性气候的理想材料。

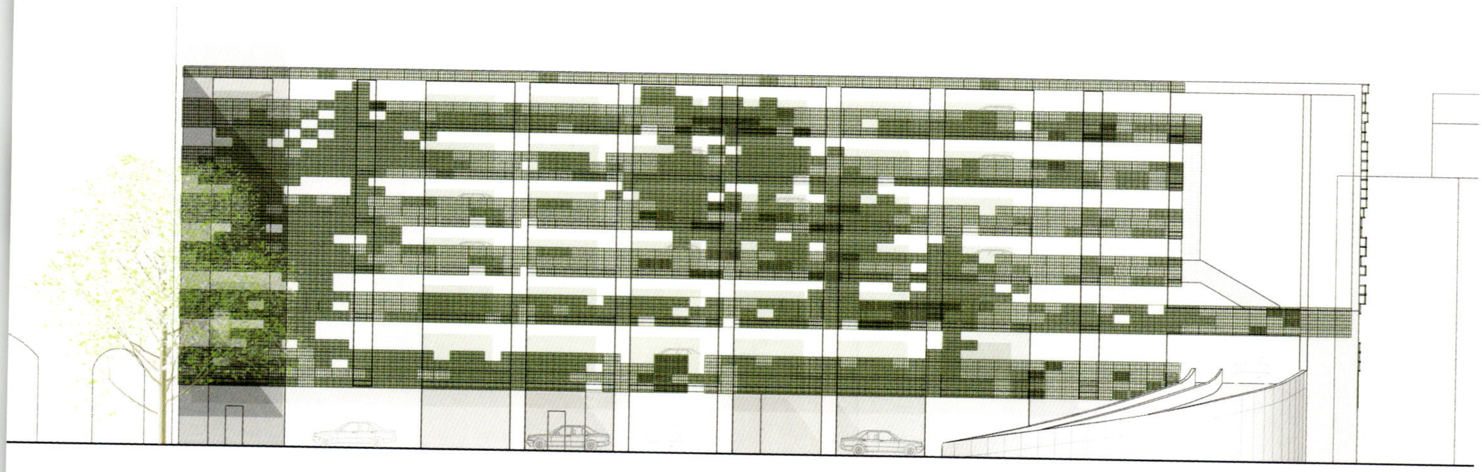

Maison de l'Eau
水质研究中心

Water-nature Study Centre, Vaulx En Velin | Design: n!studio Susanna Ferrini, Antonello Stella, with Beterem Ingénierie Rhone Alpes Centre, Les Architectes Du Paysage, Architecte Nicoletta Trasi, Atelier Bleu, Atelier Lefevre - Architectes | Team: Stefano Battaglia, Simone Bove, Francesca Fontana, Anna Hollstein, Angelo Venturi, Paolo Rossi, Andrea Cingoli | Client: Grand Park Miribel Jonage | Period: 2008

This project is focused on creating an exhibition space that emotionally and perceptively involves those who visit it, allowing them to directly experience the various phenomena of river ecosystems. The structuring idea of the proposal is that the scenogrpahy is not limited to the spaces inside the buildings, but also involves the exterior entry courtyard. It is precisely this idea of the large park that surrounds the 'eau-nature' educational centre that led us to imagine that the site itself is the object "on display". In fact, the courtyard is the first exhibition space that introduces visitors to the centre through the direct experience of the elements that make up the park, the river, the composition of the ground and its flora and fauna. The organisation of the paving in parallel strips constitutes the first level of exhibition and communication of the elements of the park. The paving strips have been designed, above all, as the space of didactic experimentation: children and students can be involved in in situ studies of these different materials. The strips are placed in continuity with the new façade of the centre, which reconstructs the long side of the courtyard with a new building. The articulation of the centre's volumes, similar to a large backdrop, is focused on interpreting the composition of the courtyard, re-proposed in the skin of the building. For example, the water pools on the ground continue up the façade as part of a wall of display cases that expose the processes of water depuration taking place behind it. The structuring idea of the project is that there is no difference or separation between the architecture of the building and the backdrop, the content of the exhibitions and the park. The themes of the park become those of the architecture, rendered communicative and 'speaking' about the themes of the large park. The content of the interior display spaces is designed in continuity with the themes exposed on the exterior façade. The glass entry volume that closes the short side of the courtyard provides access to the first level of display spaces, which features an area for

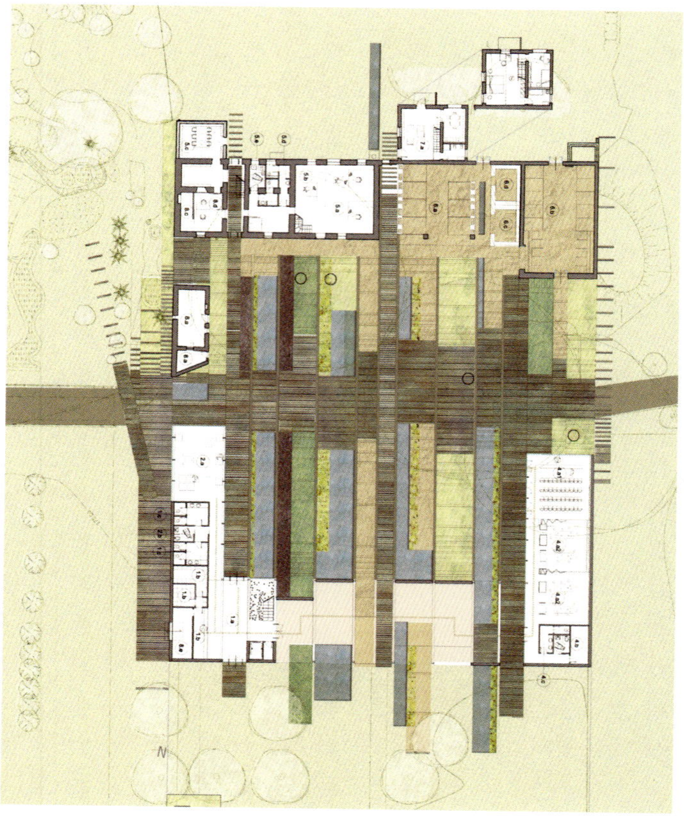

temporary shows. The permanent collection is characterised by a line that orders and organises the contents by themes. A glass strip is set flush with the floor and crosses all of the display spaces: an ideal representation of the line of the river. This glass strip in turn contains a series of elements that can be found in the park, from clay to river gravel. The second element of exhibition is the sensory tunnel, filled with further investigations of the centre's various themes, presented in display cases and projections.

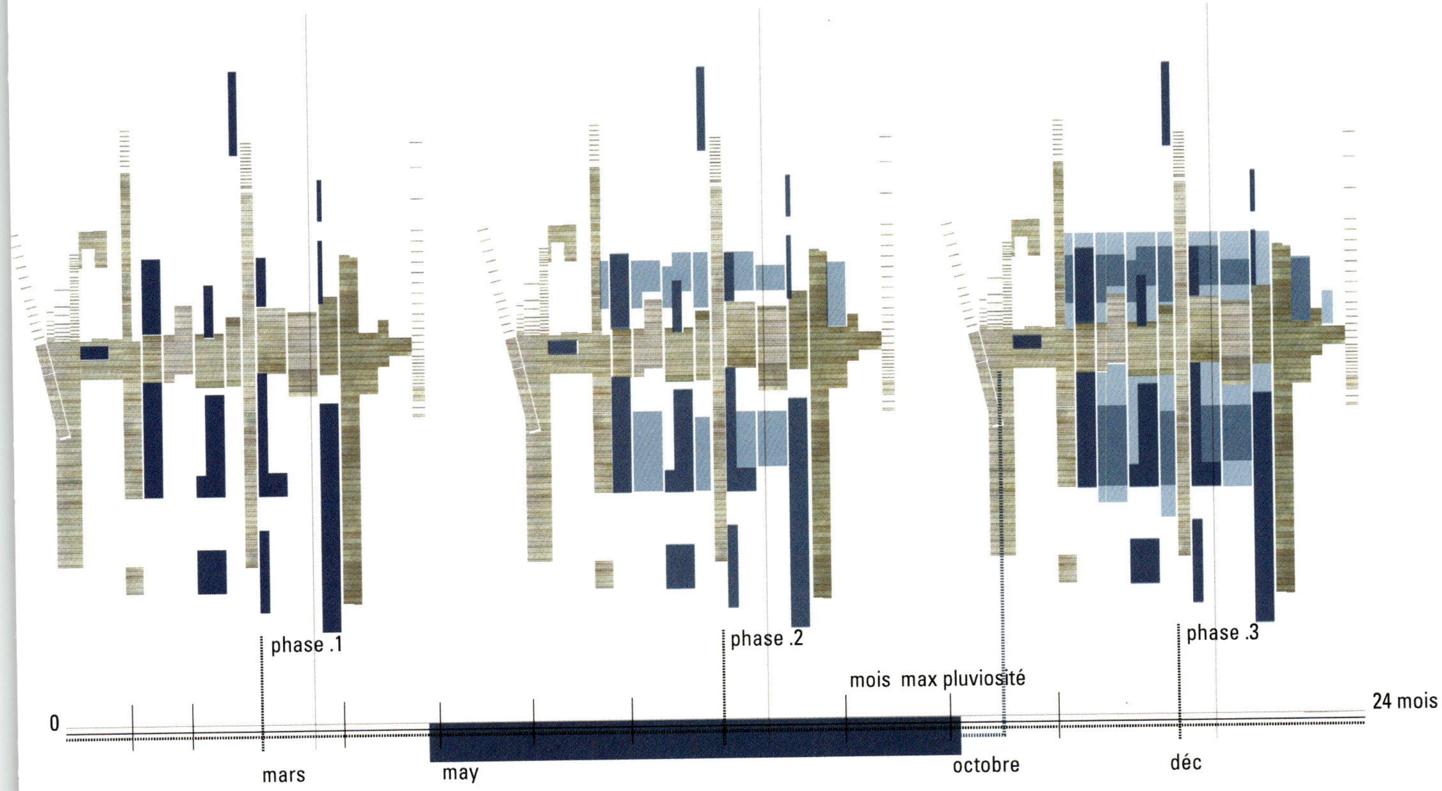

phase .1
phase .2
phase .3
mois max pluviosité
0
24 mois
mars
may
octobre
déc

本项目的建设目的集中在建造一个能够在情感和感知方面使游客融入其中的展区，使游客能够直接体验河流生态系统中的各种场景。按照我们所提出的结构构思，其配景图不仅仅局限于其建筑的内部，还包括其外部的入口庭院。正是围绕此“水质”教育中心周围建造大型公园的构思使我们能够将此项目场地本身想象成一件“展示”品。事实上，该庭院正是第一个展示区，游客们能够直接体验河流、地表组分及其植物群和动物群等各组成部分，同时将游客们引向此展览中心。其路面铺设成平行的带状结构，这种组织形式构成了第一层的展区，形成了与公园各区域之间的交流。这种带状路面的设计首先是作为一种教学实验的空间：孩子们和学生们可以在此现场学习各种不同的材料。这种带状结构同该中心的新外表面结构连为一体，在该庭院的长边一侧重新建成了一座新的建筑。此中心的连接形式就像是一个大型的背景，它集中诠释该庭院中的事物组成，重新反映在了该建筑物的表层结构中。例如，地面上的水池一直延伸到建筑物外表面结构处，成为了墙体展示柜的一部分，展示出在其背后正在进行的水净化流程。此项目的构造构思在于，此建筑物的建筑结构同其背景之间，以及所展示内容同公园本身之间没有差别和隔离结构。公园的主题变成了建筑的主题，建筑的主题与该大型公园的主题能够进

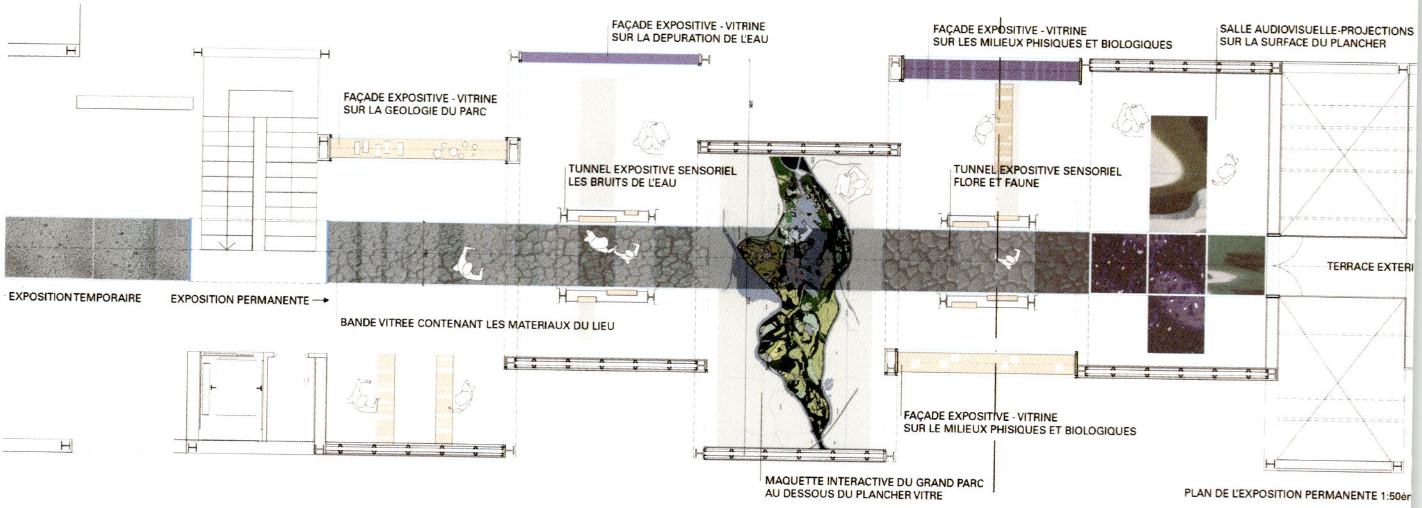

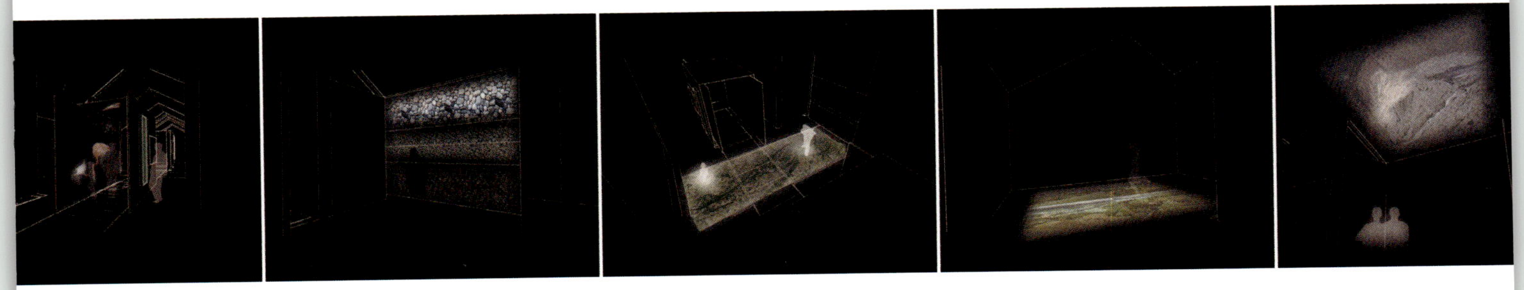

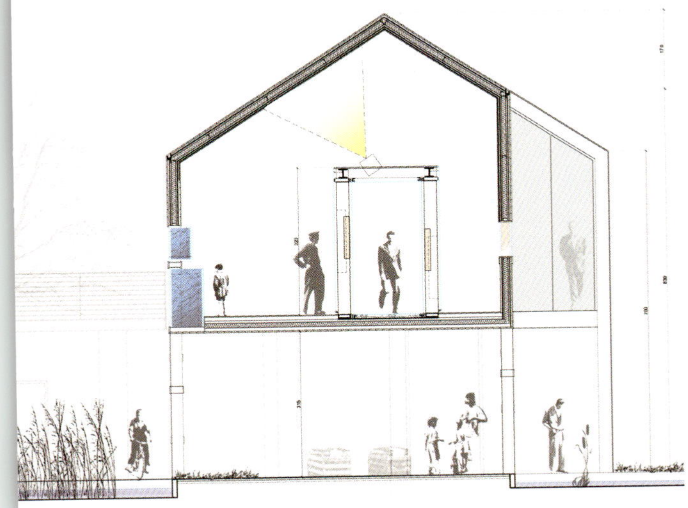

行交流，开始“诉说”。其内部展示空间的内容被设计为其外部外表面结构所显露主题的延伸。玻璃构造的入口位于该庭院短边一侧的末端，是进入第一层展示区大门，此第一层展区主要用于临时性展览。在其永久性展品区，通过各种分隔线根据主题将各种展品组织安置。一条闪烁的玻璃带状结构沿着地板穿过整个展示区——以理想的形式代表了河流的流线。这条玻璃带依次含有一系列的公园中的自然元素，从黏土到河砾石。其展示的第二种事物是一种感官隧道，充斥着对该中心各种主题的进一步研究，在展示柜和投影中展示。

Ecomuseum Rennes

雷恩生态博物馆

International Shortlisted Competition for an Agricultural Park in Bretagne - Second Phase | Design: n!studio Susanna Ferrini, Antonello Stella with David Serero | Team: Stefano Battaglia, Simone Bove, Teresa Zanaria | Collaborators: Maria Angela Corsi, Sara Ferrazzoli, Paolo Rossi Consultants: Carolina De Camillis, Riccardo Fibbi Lighting Design | Engineering: Beterem Ingenierie | Client: Communaute d'Agglomeration Rennes Metropole | Period: 2006 | Model: Marco Galofaro | Award: 2nd Place

ecobuilding

The theme of this project is the extension of the current ecomuseum in rennes in two new volumes located alongside the existing and current entrance to the farm, inserted within the site based on the same logic that generated the buildings of the local historical farms: the addition of successive new buildings around a void, central space. The rennes ecomuseum, created inside the 'bintinais' farm, serves as the custodian of the culture and memory of this territory, seeking to reveal it to visitors through the conservation of flora and fauna from the region. Our project, even while searching for ties and inspiration within the landscape, does not limit itself to the use of traditional architecture, but rather offers a new identity, inevitably contemporary, to the ecomuseum. Similar to a natural process, the ground rises and separates into three slopes, creating the temporary exhibition hall, the entry foyer and defining the new façade of the existing building. The façade and roof of these buildings are treated as a unitary, green surface that rises out of the earth and extends the landscape, while the western façade features large glazed surfaces that open towards the fields surrounding the ecomuseum. The façade/roof, partially covered in gravel, partially covered in plant material of varying density, is inspired by the profile of an apple tree: the circular motifs define the natural "vases" of earth that contain examples of local flora. This roof also allows for the absorption of rain water and increases the thermal inertia of the building. The structure of the large hall is composed of steel trusses that permit a column-free space for temporary exhibitions and the glazing can be opened, projecting the space into the landscape.

本项目的主题是将目前的雷恩生态博物馆进行扩建，在农场目前现有的入口旁边修建两座新的建筑体，采用与当地历史性农场建筑相同的建筑逻辑将它们嵌插入当地场地之中，即在空旷的中心区域增添新的建筑物。雷恩生态博物馆修建在Bintinais农场之内，它的作用是作为当地文化和历史回忆的场所，通过保护该地域的植物群和动物群来将其展示给游客。在我们的项目中，即使是在其地貌风景中寻找关联之处和灵感的过程中，也并未仅是限制在使用其传统的建筑，而是为该生态博物馆提供了一项新的特性，不可避免地，这项新的特性是属于当代的。这些建筑的正面表面和屋顶是从地面升起的整体型绿色表面，进一步延伸了其地貌风景，而西边的正面则以大型光滑表面为特色，直接通向该生态博物馆周围的田野。其正面和屋顶部分覆以砂砾层，部分覆以不同密度的植物材料，这种灵感来自苹果树的构造：其环形的样式形成了由土地构成的天然"花瓶"，花瓶中栽种着当地植物群中的精品。这种屋顶还能够吸收雨水，增加该建筑的热惰性。其大型展厅的结构由钢制支架构成，为临时性展览提供了没有支柱的大型展区，而且其玻璃窗可以打开，使其空间延伸至周围地貌风景之中。

EXPOSITIONS TEMPORAIRE
PERIMETRE SYSTEME D'OCCULTATION DE LA SALLE D'EXPOSITION
HALL
ACCUEIL
ADMINISTRATION
LEGENDE
PARCOURS PU
PARCOURS PE
ACCES LIVRAIS
TABLEAU DE SURFACES
1. EXPOSITION TEMPORAI
1.1 SALLE D'EXPOSITION
1.2 SALLE DE PREPARATIC
1.3 STOKAGE
2. ESPACES D'ACCUEIL
2.1HALL D'ENTREE
2.1.1 ESPACE VENTE
2.1.2 ESPACE REPOS
2.2BILLETERIE
2.2.1 VESTIAIRE
2.3 LOCAL TECHNIQUE
2.4 STOKAGE PAPETERIE
2.5 CENTRE DE DOCUMEN
2.6 SANITAIRE DU PUBLIC
2.7 MAINTENANCE
2.8 LOCAL INFORMATIQUE
2.9 CHAUFFAGE
2.10 LOCAL DECHET
2.11 AUVENT PARKING
2.12 AUVENT PARC
3. LOCAUX ADMINISTRATI
3.1BUREAU CONSERVATEU
3.2BUREAU REDACTEUR
3.3 SECRETARIAT 1
3.4 SECRETARIAT 2
3.5 BUREAU POLYVALENT
3.6 SALLE DE REUNION
3.7 REPROGRAPHIE
3.8 ARCHIVES
3.9 SALLE DU PERSONEL
3.10 SANITAIRES DU PERS

ECOMUSEE
DU PAYS
DE RENNES

Museum of Water

水博物馆

International Competition for the Parc Du Massif Des Bauges, Cheran - Second Phase | Design: n!studio Susanna Ferrini, Antonello Stella with Bernard Bellion Jourdan, Nicoletta Trasi | Team: Simone Bove, Gabriele Misso, Tiziana Monti, Loris Rossi, Giulia Scaglietta | Consultants: Carolina De Camillis, Riccardo Fibbi Lighting Design | Engineering: Beterem Ingenierie | Client: Parc Naturel Regional du Massif des Bauges | Period: 2005 | Award: Joint 1st Place

landscape

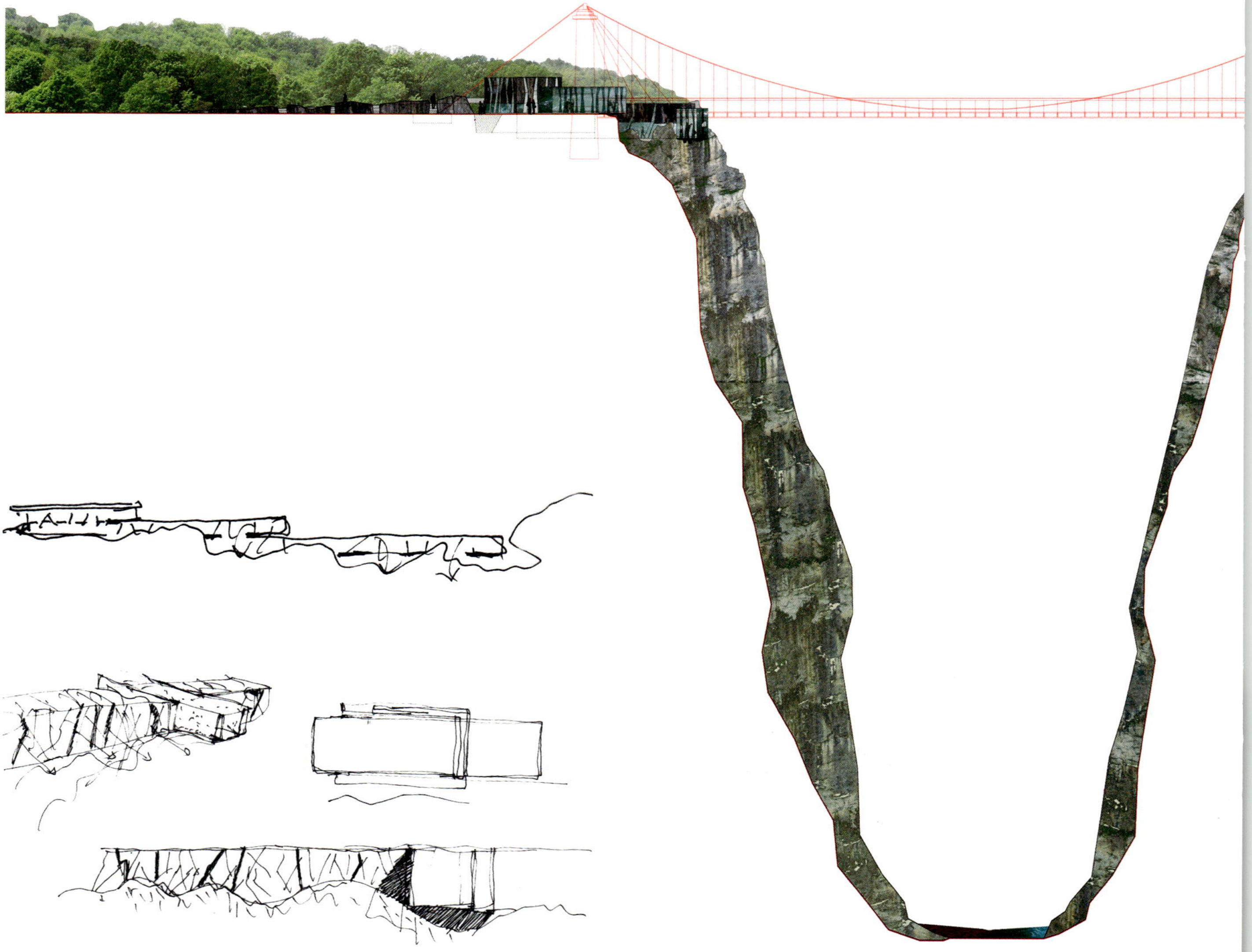

A road and a large, sloping plane provide access to the edge of the abyss, along the 120 meter high rock wall that faces the cheran torrent, in the parc du massif des bauges. The project intended, on the one hand, to underline the technical prowess of the suspended bridge designed by arnoudin, which gracefully extends over the cheran gorge and, on the other, to free up the view of the surrounding rocky landscape. Our intention was that of putting on display, without any rhetoric, natural, mineral and vegetal elements, in the search for a strong integration between architectural form and the landscape. It is a project that adapts to the terrain, one could almost call it a work of topography, with a minimal volumetric impact. The maison del la rivière cheran is inserted into the landscape in continuity with the cliff: the exterior walls are composed of panels of reinforced concrete, whose design recalls the surface of the adjacent rock walls; the roof is a grassy plane that reduces the impact of the building when seen from atop the rock massive. The façade features vertical openings that recall geological fractures, created using glass panels. The glass introduces transparency, in turn a reference to water, crystal and ice. The volumes rest on the cliff and the fractures-windows continue like incisions in the grassy roof, widening into large skylights that become glass volumes in the interior display spaces. The interior floor is modelled in sloping surfaces set on rocky platforms, once again in continuity with the exterior cliff.

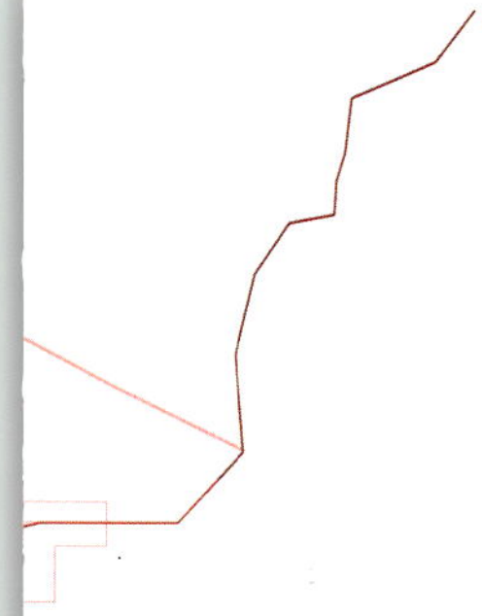

Lunettes

Maison de la rivière

在Parc du Massif des Bauges，沿着高120米、濒临Cheran湍流的岩壁，我们可以通过一条道路和一个大型倾斜平面抵达深渊的边缘。所预期进行的项目在一方面是要强调由Arnoudin设计的优雅地横跨在Cheran峡谷上的悬浮桥的技术能力，另一方面要释放出周围岩石景观的美景。我们的目的是在不采用任何矫饰的、天然的、矿物及植物元素的情况下，展示对建筑形式和地貌景观之间强力结合的追求。这是一项适应地形而建造的项目，我们几乎可以称之为一种地形学作品，建筑体对地形的影响非常之小。Maison del la Rivière Cheran沿着悬崖的走势插入到了地貌景观之中：其外墙由钢筋混凝土板构成，其设计重现了其邻近岩壁的表面形态；其屋顶是一个绿草茵茵的平台，使该建筑物与地貌环境相融合。其外表面具有由玻璃板建造的独具特色的竖直开口，重现了天然地质学裂缝的外观。玻璃所具有的透明度同水、晶体和冰等物质有着相似性。该建筑体位于悬崖顶上，其裂缝式窗子就像各种切口一样一直延伸到布满青草的屋顶上，逐步变宽而形成大形的天窗，在其内部展示区又变成了玻璃建筑体的形式。其内部地面设置在石头平台上，形成倾斜的表面，同样延续了外部悬崖的走势。

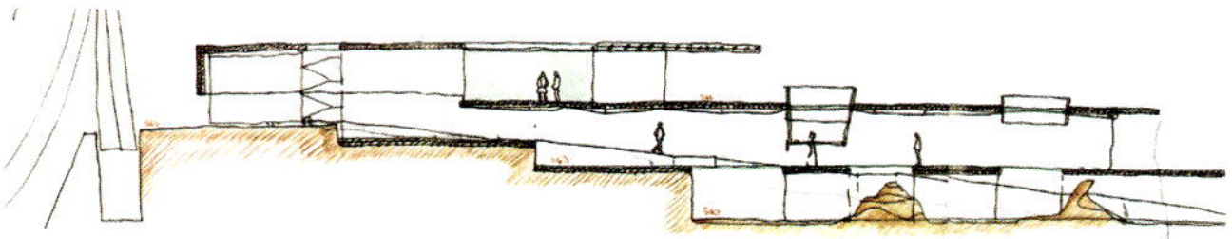

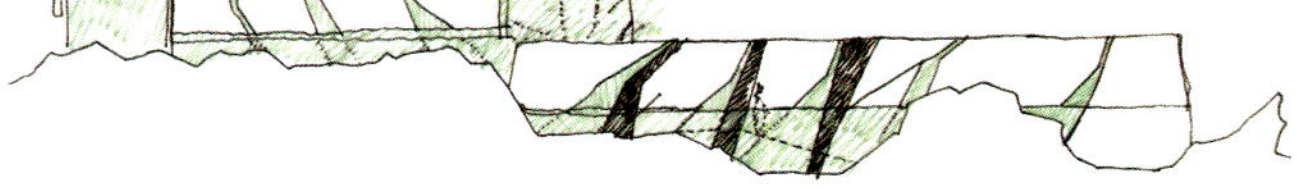

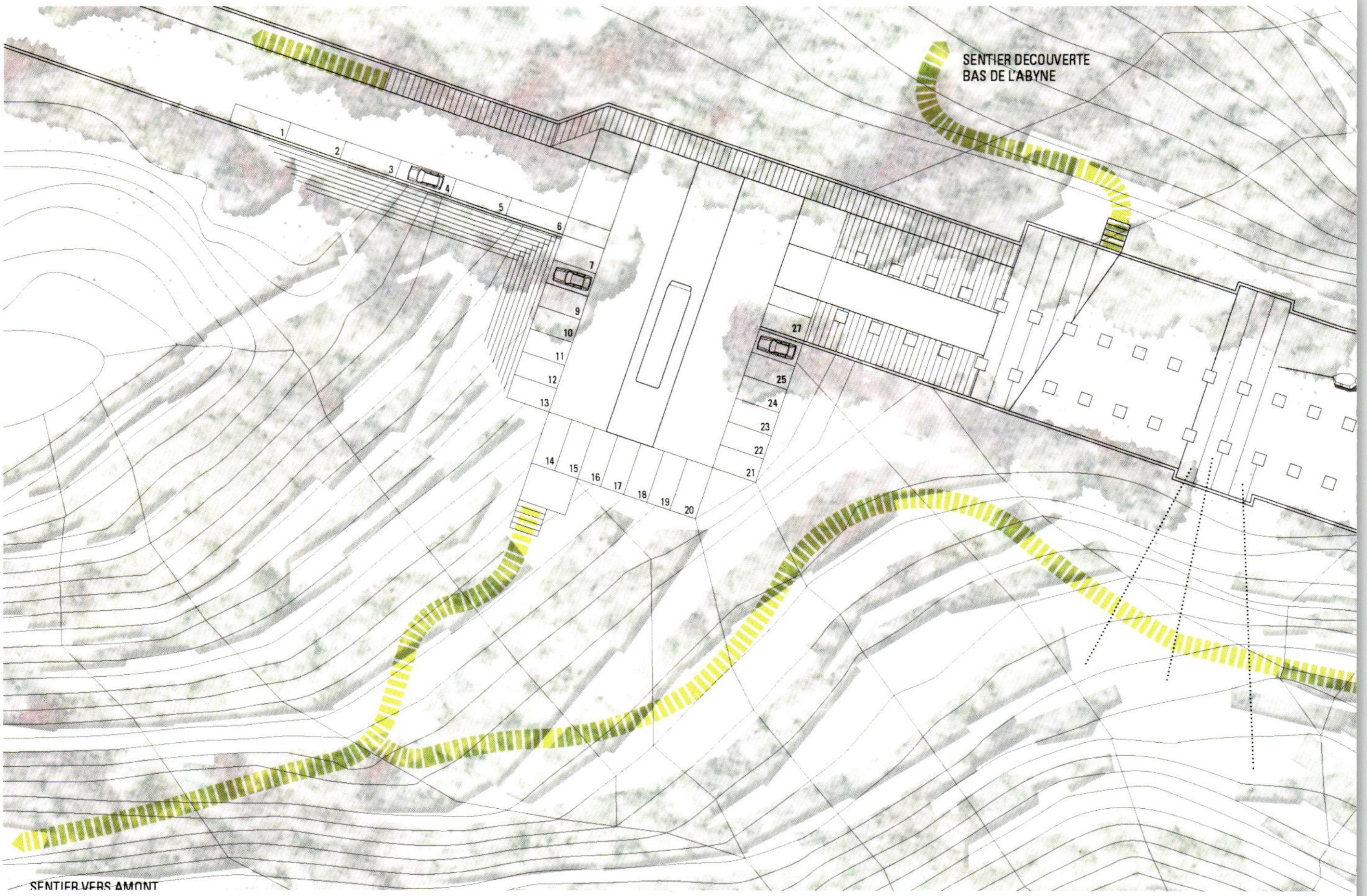
SENTIER DECOUVERTE
BAS DE L'ABYNE
1
2
3
4
5
6
7
9
10
11
12
13
14
15
16
17
18
19
20
21
22
23
24
25
27
SENTIER VERS AMONT

BELVEDERE
EE

Froissy Archaeological Museum

FROISSY考古博物馆

International Shortlisted Competition for the New Gallic-Roman Museum, Froissy - Second Phase | Design: n!studio Susanna Ferrini, Antonello Stella with David Serero | Team: Simone Bove , Stefano Battaglia, Mariangela Corsi, Tiziana Monti, Paolo Rossi, Giulia Scaglietta, Teresa Zanaria Consultants: Carolina De Camillis, Riccardo Fibbi Lighting Design | Engineering: Betom, 3ti Client: Communaute Des Communes Des Vallees De La Breche Et De La Noyre | Period: 2006-2007 | Model: Marco Galofaro | Award: Winning Project

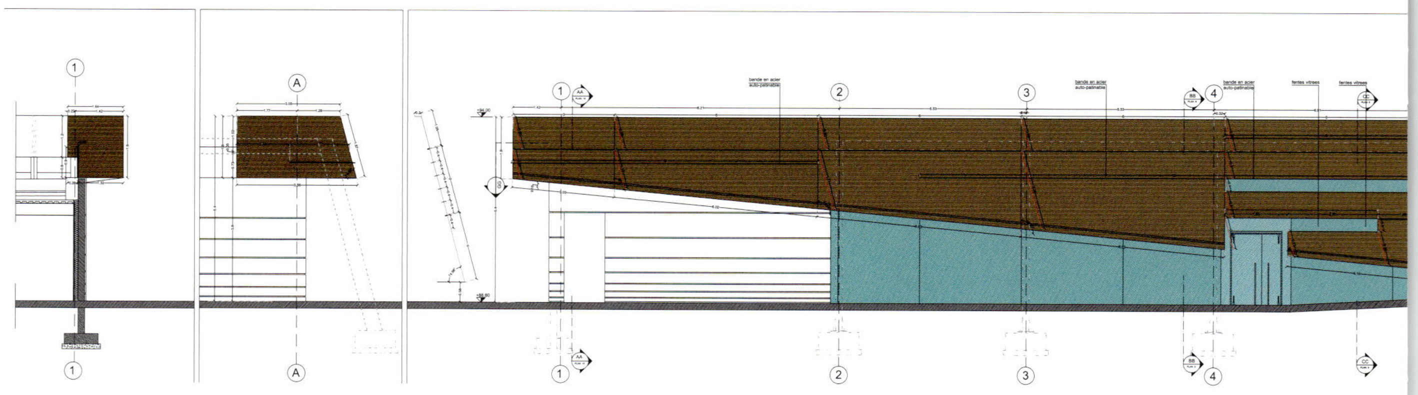

bande en acier
auto-patinable
bande en acier
auto-patinable
bande en acier
auto-patinable
fentes vitrees
fentes vitrees

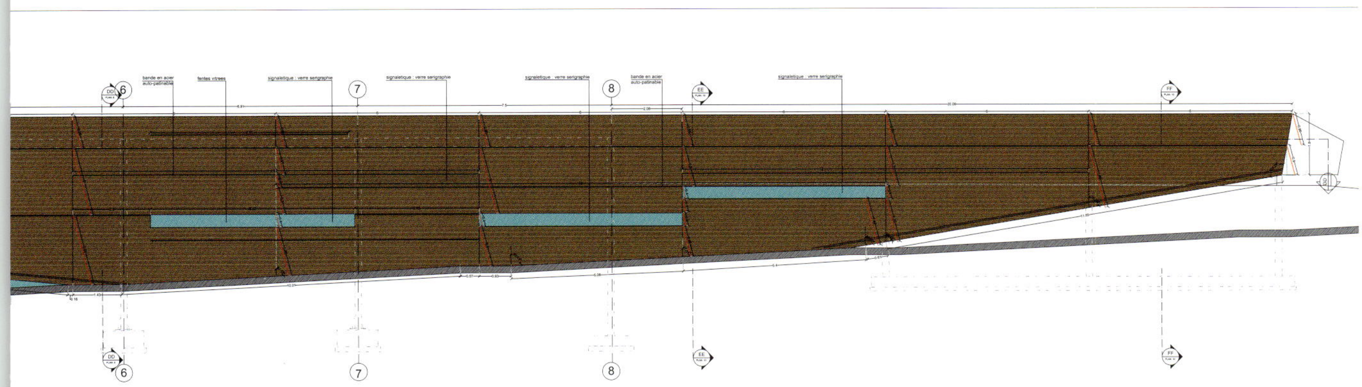
bande en acier
auto-patinable
fentes vitrees
signaletique : verre serigraphie
signaletique : verre serigraphie
signaletique : verre serigraphie
bande en acier
auto-patinable
signaletique : verre serigraphie
6
7
8
DD
EE

The project set into the earth, almost invisible from the gallic-roman amphitheatre. It is designed as an extension of the landscape, an insertion within the chain of valleys and depressions that the picardie topography forms along the horizon. The museum is a sign, excavated into the earth, a building that is not defined by its presence or its materiality but, on the contrary, by its absence. It is through a play of excavations and openings that the museum attempts, like the archaeological site that surrounds it, to reveal the invisible. The museum allows itself to be discovered by visitors through movements, walking along a cor-ten steel ramp from the level of the road towards the entrance to the museum, in the lower area. This ramp, positioned along the axis of the future path that will lead to the amphitheatre, almost slides the visitor into the depth of the earth, displaying its various layers, passing from the natural, the existing landscape, into the artificial, that of the museum, arriving at the layers of archaeology, which sink even deeper. The museum is also simultaneously the rising of the ground plane, allowing for the opening of a cut in the landscape to create the entrance and bring natural light into its interior. Along the access ramp, the only façade of the museum is a blade of cor-ten steel that contains the ramp, a retaining wall and the display space of the museum. Oxidation naturally gives cor-ten steel its own protective layer against humidity and corrosion. Its copper hues render it almost organic as its patina and colouring change over time.

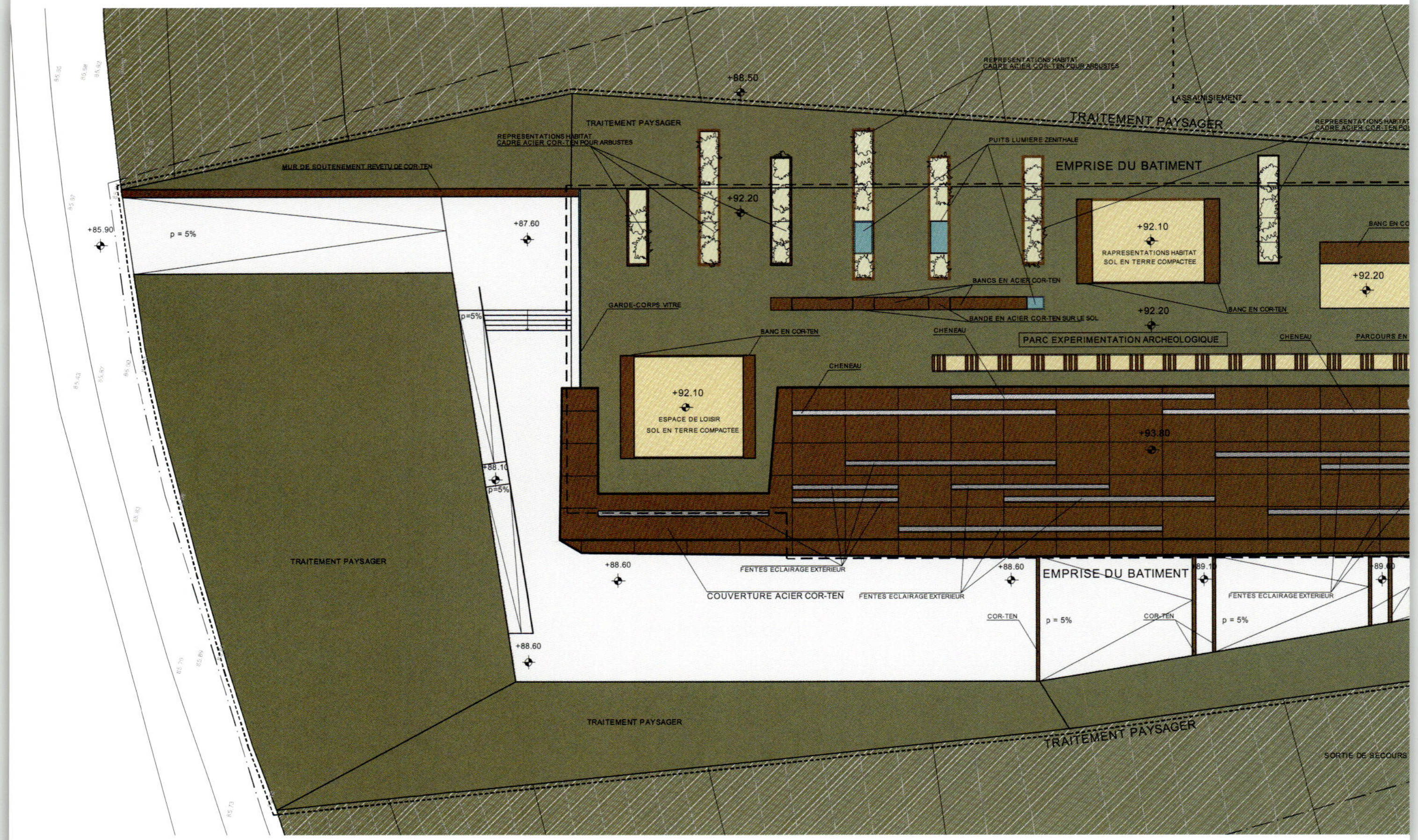
+88.50
REPRESENTATIONS HABITAT
CADRE ACIER COR-TEN POUR ARBUSTES
ASSAINISSEMENT
TRAITEMENT PAYSAGER
TRAITEMENT PAYSAGER
REPRESENTATIONS HABITAT
CADRE ACIER COR-TEN POUR ARBUSTES
PUITS LUMIERE ZENITHALE
EMPRISE DU BATIMENT
MUR DE SOUTENEMENT REVETU DE COR-TEN
+85.90
p = 5%
+87.60
+92.20
+92.10
RAPRESENTATIONS HABITAT
SOL EN TERRE COMPACTEE
BANC EN COR-TEN
+92.20
BANCS EN ACIER COR-TEN
BANC EN COR-TEN
BANDE EN ACIER COR-TEN SUR LE SOL
+92.20
GARDE-CORPS VITRE
p=5%
BANC EN COR-TEN
CHENEAU
PARC EXPERIMENTATION ARCHEOLOGIQUE
CHENEAU
PARCOURS EN
CHENEAU
+92.10
ESPACE DE LOISIR
SOL EN TERRE COMPACTEE
+93.80
+88.10
p=5%
TRAITEMENT PAYSAGER
+88.60
FENTES ECLAIRAGE EXTERIEUR
+88.60
EMPRISE DU BATIMENT
COUVERTURE ACIER COR-TEN
FENTES ECLAIRAGE EXTERIEUR
FENTES ECLAIRAGE EXTERIEUR
COR-TEN
p = 5%
COR-TEN
p = 5%
+88.60
TRAITEMENT PAYSAGER
TRAITEMENT PAYSAGER
SORTIE DE SECOURS

+93.30
MUSEE-CENTRE DE RESSOURCES ARCHEOLOGIQUES
+88.50
+88.50

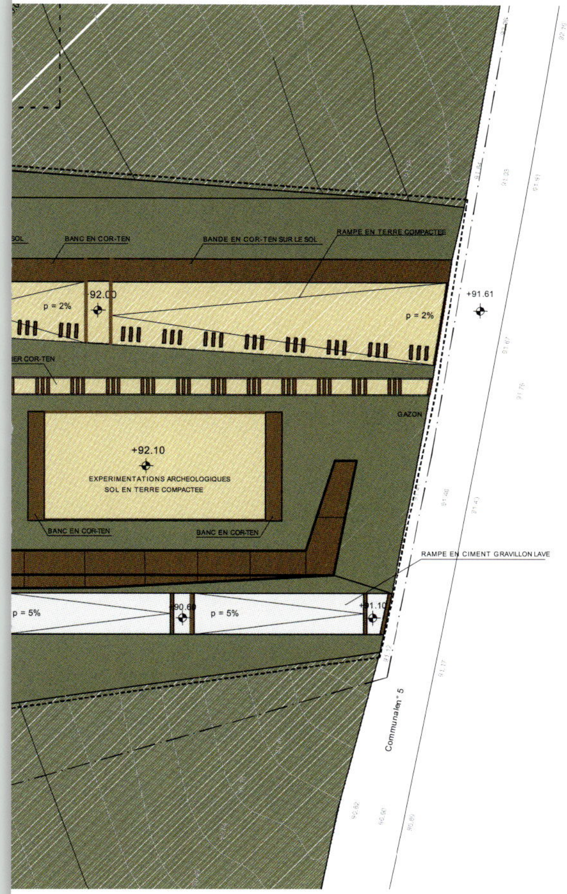

本项目设置在地下，在Gallic–Roman的比赛场中几乎看不到。其设计就像地貌景观的延伸，插入到了峡谷链以及Picardie地形沿地平线形成的凹陷中。此博物馆成为了一个标志，凹入地下，是一座并不是因为其存在或其物质性，而是因为其不存在而具有意义的建筑。该博物馆尝试着像其周围的考古区一样，通过空穴和开口的设置，将其隐匿的内涵展示给世人。游客可以沿着公路行走至位于较低区域的博物馆入口的低合金高强钢坡道上，在过程中不断发现其奥妙之处。该坡道沿着计划建造的通往圆形露天剧场的通道的轴线建造，几乎使游客滑入了地层的深处。沿途可经过不同的层次，从天然存在的地貌风景层进入博物馆的人工建造区，最后抵达更深的考古发掘层。该博物馆在地貌风景中打开了一个豁口，形成了博物馆的入口，并使自然光线能够进入其内部。沿着入口坡道，博物馆的唯一正立面就是容纳了该坡道的低合金高强钢突出体，也是该博物馆的档土墙和展示区。通过氧化作用低合金高强钢形成了防湿和抗腐蚀的防护层，像有机物一样，其铜锈和色彩会随时间而变化。

MUSEE-RESERVE
ARCHEOLOGIQU

+93.20
+90.30
+91.00
+88.50

MUSEE-RESERVE ARCHEOLOGIQUE FROISSY

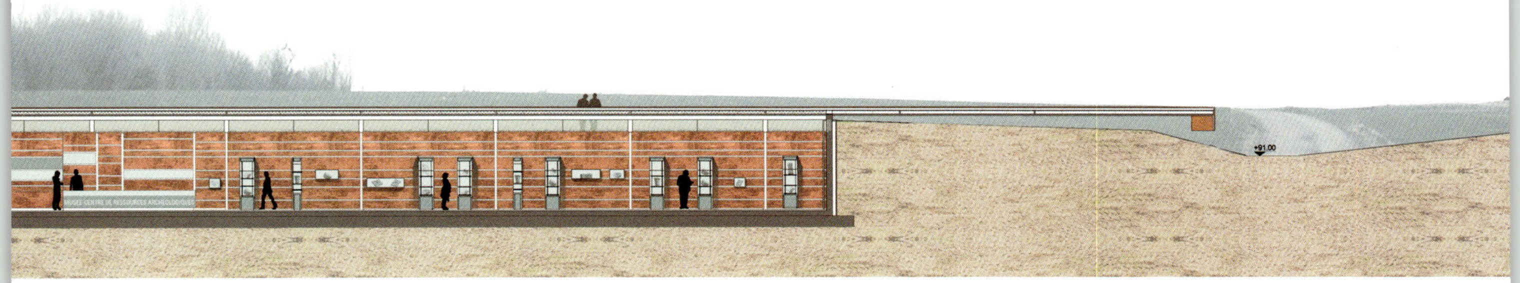
+91.00

Maison de la Lumiére

灯光博物馆

Restricted International Competition for the Construction of the Museum of Light, Gap France | Design: n!studio Susanna Ferrini, Antonello Stella | Team: Simone Bove, Teresa Zanaria, Maria Angela Corsi, Sara Ferazzoli | Consultants: Bernard Bellion-Jourdan | Engineering: Etech Ingénierie, Betech Ingénierie | Client: Conseil General Amenagement et Developpment, Gap | Period: 2006 | Award: 2nd Place

l'univers commence a ressembler plus a
la lumierere est corpusculaire
soleil, ce n'est pas uniquement de la chal
mais la lumire est totalment fantastique
MAISON DE LA LUMIERE

Starting point for the project "house of light" is the landscape: a natural landscape which merges into an artificial landscape, a path that is crossing over the natural platform of the site «du Chevalet» of Aspre Sur Buech from west to east, tracing a new line of horizon. The new project contains space for exhibitions on the theme light in all its different forms and phenomenon's, a meeting room and a room for projections. The project aims to integrate the existing building "the pyramid", the actual headquarter of the light exhibition. The visitor, who will start from the western side of the area, follows the line of light, which engraves and modulate the ground, to reach the entrance hall of the museum. In the beginning the path appears only as a sign of pavement, but along the way it becomes always more and more dense and compact. In the end the path lifts up to become first surface and finally form. Unlike usual buildings, the museum develops like a tunnel, like the perception of light that becomes matter, a path that integrates architecture, as well as experience and visitors in a peculiar system of discovery.

此"灯光屋"项目的设计起点是风景：一种融入人工风景中的自然景观，一条从西到东横跨位于Aspre sur Buech 的"du Chevalet"场地的自然通道形成了一道新的地平线。此新项目包括表现为各种形式和情境的灯光主题的展示区、会议室和投影室。此项目旨在将现有建筑"金字塔"融入其中，该建筑是此灯光展示区的主体。游客们将从西侧开始其行程，随着灯饰的指引，抵达博物馆的入口大厅。此通道的起点只是一条普通的人行道，但是沿着通道前进将变得越来越丰富多彩。在通道的末端最终形成了博物馆的第一层表面。与一般的建筑不同，此博物馆就像一条隧道，对灯光的感悟形成实质化的过程，将游客带入了一条独特的发现体系当中。

LE SOLEIL, CE N'EST PAS UNIQUEMENT DE LA CHALEUR, SA LUMIERE DEVIENT ELECTRICITE...
ENERGIE SOLAIRE

3.95
3.80
+0.00
+0.00
+1.00
+1.00
+1.00

MAISON DE LA LUMIERE
LA ROUTE DE LA LUMIERE
Grenoble
E=mc2

MAISON DE LA LUMIERE

ENERGIE SOLAIRE

MAISON DE LA LUMIERE

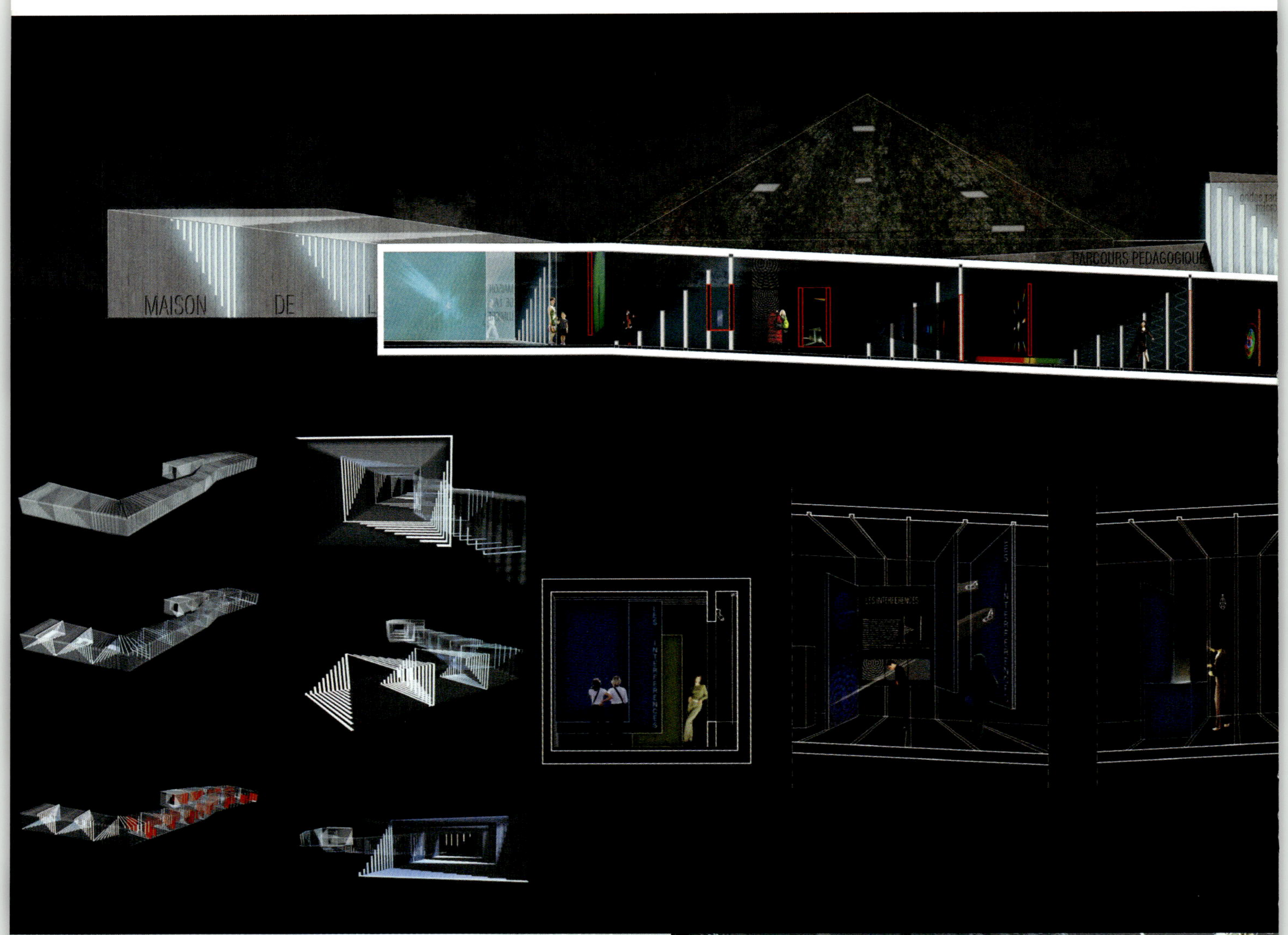
MAISON
DE
PARCOURS PEDAGOGIQUE

SITE DE LA LUMIERE

Descartes
Galileo
Newton

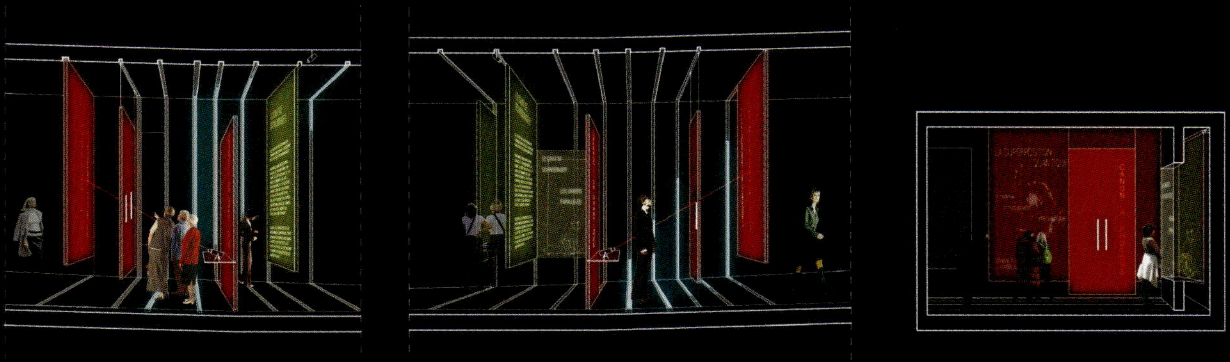

LES INTERFERENCES
SUPERPOSITION QUANTIQUE
LA MECANIQUE QUANTIQUE
LE SITE DE LA LUMIERE ET DE L'AIR

Piezometric Water Tower

Piezometric水塔

Water Pumping Station in the Po Valley, Codigoro | Design: n!studio Susanna Ferrini Antonello Stella | Team: Valerio Dedivitiis, Anna Cornaro, Franco Lancioni Jutta Riehle Collaborators: Stefano Battaglia, Giulia Mazzarelli, Loris Rossi | Engineering: Ing. Fabrizio Brunetti | Consultants: Carolina De Camillis, Riccardo Fibbi Lighting Design Pmg Group Budgets And Specifications | Client: Consorzio di Bonifica Del Polesine Ferrarese, Ing. Matteo Giari | Period: 20 Currently in the Tendering Phase

landscape

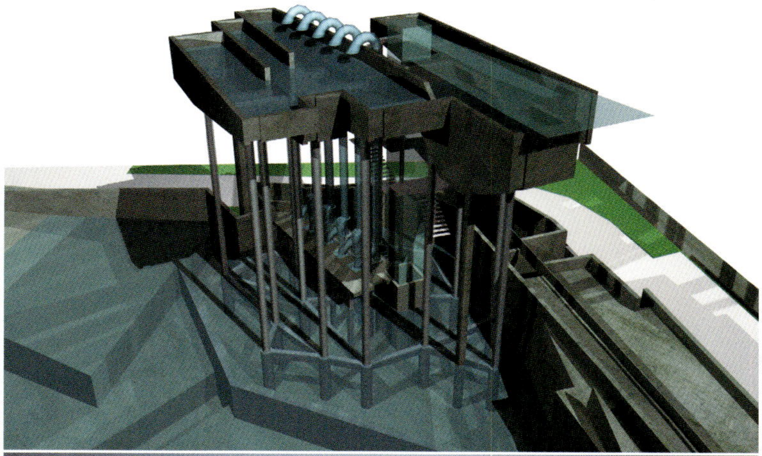

The particularity of this piezometric tower is the fact that it is used as a water tower for irrigation. The difference with common water towers can be found in its proportions: the first are as tall and slender as the second are low and with a diameter, in this case (18 m), equal to their height. The environmental insertion of this object is sought out, not through its impossible camouflaging within nature and/or 'artificial' and massive plantings of trees, but rather through 'actions' of reciprocal transformation. This piece of infrastructure tends to modify its consolidated typology, perhaps more the result of habit than any strict technical requirements, relating to the characteristics of the environment and modifying aspects of the landscape, which in turn 'contaminates' the built work. Based on this premise, the proposal can be described through 'design actions' that, through the sequence described, have guided the overall definition of the project. The first operation was that of redefining the ground plane upon which the pumping structure sits and, in particular: creating sloping earth berms, covered with a geo-textile, to house the mechanical equipment related to the functioning of the tower, and transforming the basin-reservoir used to collect water into a 'reflecting pool'. The second operation deals with the vertical supports for the reservoir, the water piping, the stair and the elevator. The primary idea is that these elements create a sort of gigantic and 'artificial' cane thicket, upon which the elevated water reservoir rests, and which sit in the reflecting pool of the water basin-reservoir. The third operation was perhaps the most unusual in this type of project: decomposing the volume of the elevated reservoir, allowing the sun's rays to penetrate its interior.

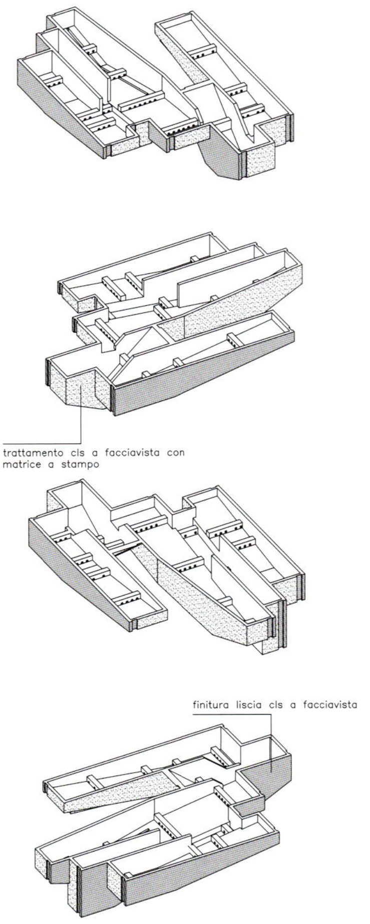

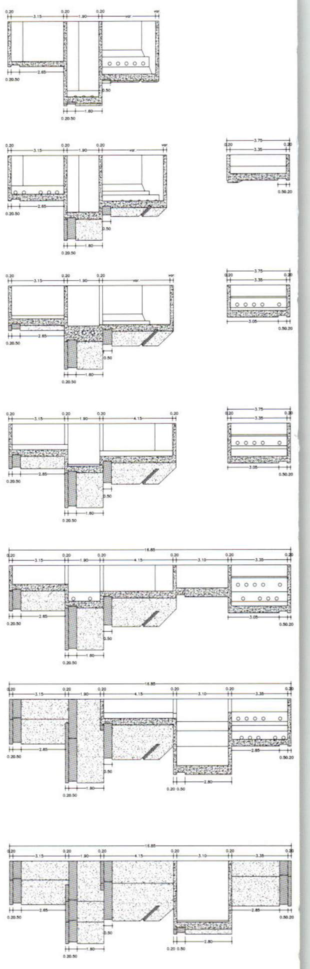

Piezometric水塔的特殊性在于，它是一座用于灌溉的水塔。在其比例方面与平常的水塔有不同之处：常用水塔高而细，而灌溉水塔低且直径大，本水塔的直径同其高度相等（18米）。本项目是符合环境保护要求的设计，不是将其置于“人工”环境和大量栽植的树木之中的方式，而是通过相互转化“措施”而实现的。此项基础设施有更改其固定地型的趋势，这也许更可能是习惯而不是任何严格技术要求的结果，它同环境的特色达成了一定关系并改变着风景地貌，而这些反过来也会“污染”所建成的建筑。在这一前提的基础上，可以通过“设计措施”来描述所提出的方案，而这些“设计措施”通过所述顺序指导了对整个项目的界定。所采取的第一项措施就是重新界定水泵建筑所在场地的地面，具体而言就是：建造覆盖有土工织物的倾斜土路，以容纳与水塔功能相关的机械设备，并将用于集水的蓄水池改造成为一种“反射池”。第二项措施对该蓄水池的竖直支撑、水的管路、楼梯和电梯进行了处理。其主要概念就是通过这些部件建造一种巨大的“人工”藤茎密集屏障，被升高的蓄水池就位于其上，而它又位于蓄水池的反射池中。第三项措施可能是这种项目中最不常用的一种措施：将对被提升起的水池进行分割，使阳光渗入其内部。

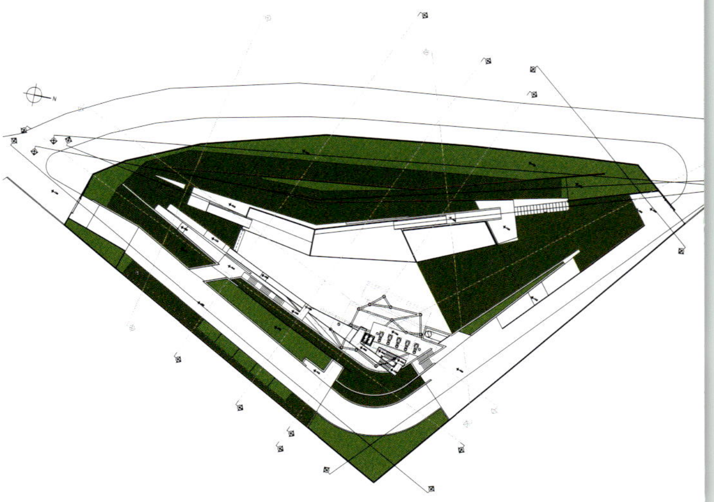

Nazzano River Museum

Nazzano河流博物馆

ovation of the Savelli Castle and Design of the New River Museum In Nazzano, Rome | Design: n!studio Susanna Ferrini, Piero Fumo, Paolo nti, Roberto Rosati, Laura Rossi, Davide Sani, Antonello Stella | Team: Dante Cecili, Francesca Federici, Birgit Dittmar, Paolo D'addato, Mauro zarelli, Paola Pozzorini, Lorenza Scalia | Consultants: Carolina De Camillis, Riccardo Fibbi Lighting Design, Alessandra Montania, Lavinia Cozza phics, Umberto Pessolano Nature Section Designer, Federico Marazzi Anthropological-Historical Section Design, Roberto Rosati Budgets Specifications Engineering: Ing. Vincenzo Pennacchi | Client: Nazzano Municipality | Contractor: Ragnini, La Corte Exhibition Setting | Period: 6-2000 | Awards: Winner Inarch Lazio 2001, Honourable Mention Gold Metal Triennale of Milan 2006, Winner Amate L'Architettura 2008

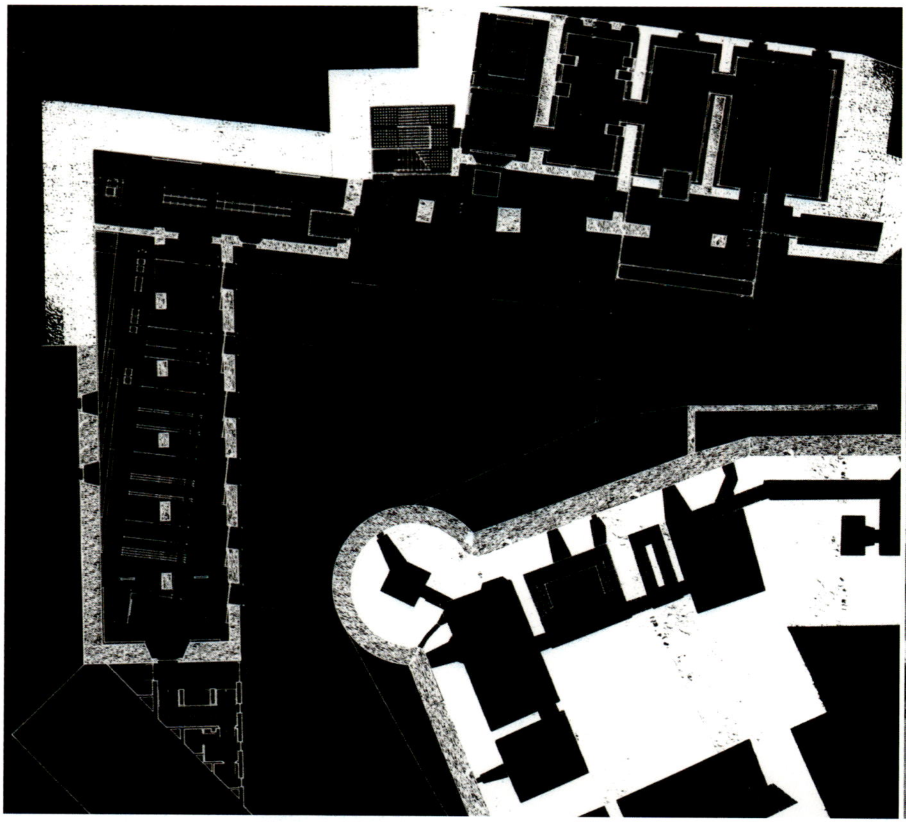

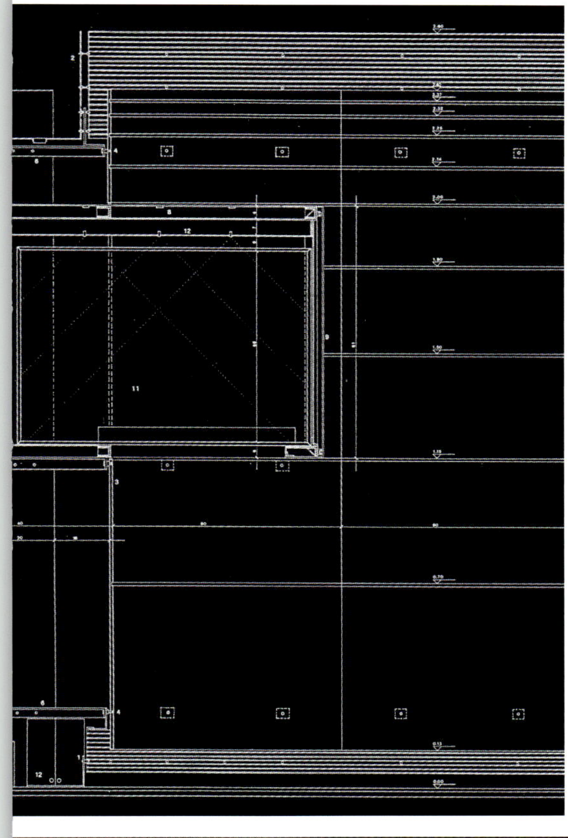

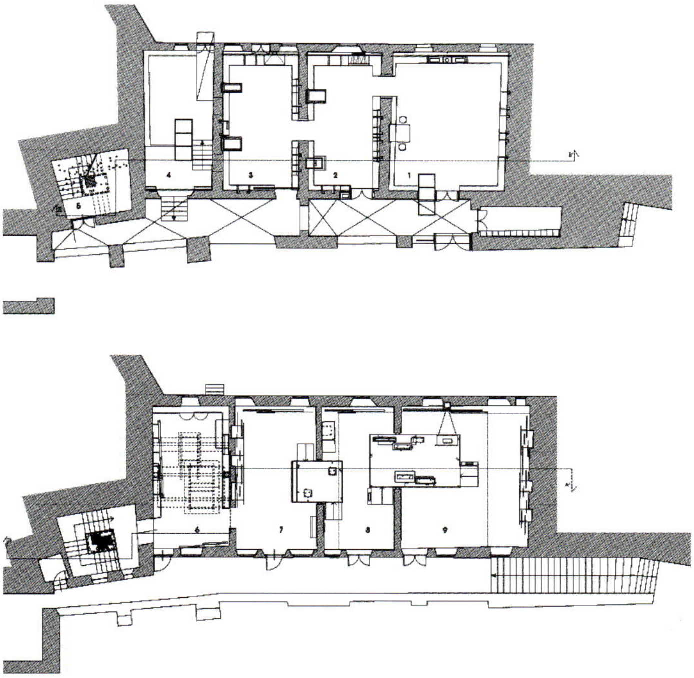

Tra i ciottoli alcune larve di plecotteri e efemerotteri attendono le possibili prede (x10)

The Science Centre and Museum is housed in part of the monumental complex of the Rocca di Nazzano, composed of the 13th century Savelli Castle and its ancient 16th to 17th century annexes.
The building, located in the public square that provides access to the fortress, contains the museum's temporary exhibition spaces and conference rooms, the Fossil Museum and the permanent section of the River Museum.
The museum is seen as a space of interest and differentiation, where the visitor can actively experience the complexity of the Tiber River ecosystem. In fact, the choice to use glass as the primary material for the project recalls the colour, reflections and transparency of the water of the river to which the museum is dedicated.
Our aim was to eliminate, as much as possible, the distance between the visitor and the objects on display, ensuring that the display cases become elements of great interest for the observer-visitor: the display cases are not simple 'containers' for objects, but real 'environments' that one can enter and, for example, perceive the sounds of the river environment or walk directly atop an idealised river-bed.
On the first floor of the museum the display elements are

BORATORIO
SSIDERMICO
LABORATORIO
NATURALISTICO

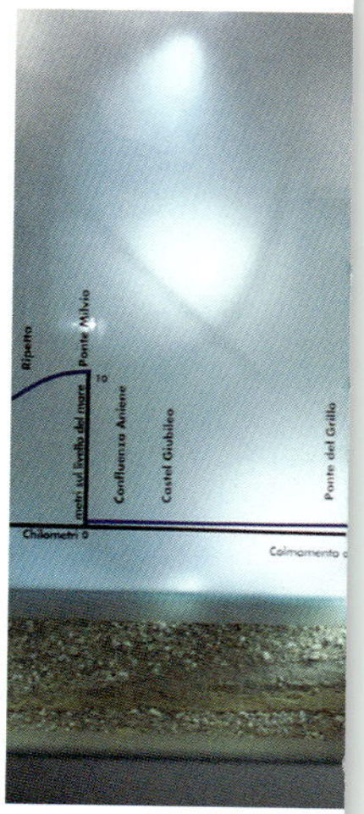
Ripetta
Ponte Milvio
metri sul livello del mare
10
Confluenza Aniene
Castel Giubileo
Ponte del Grillo
Chilometri 0

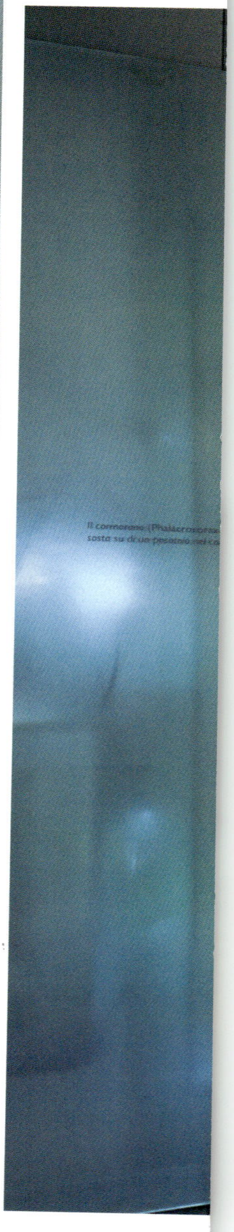

composed of a continuous 'skin' in acid-etched glass, 2.4 meter high glass wall containing information panels, dioramas, back-lit projections and cantilevered display cases: the visitor reads the texts and observes images of the river's ecosystem on the pages of a 'three-dimensional' book.

On the second floor two 'environment-display cases' allow the visitor to enter inside and visually and acoustically experience the elements that characterise the river environment.

Even the stair that connects the two levels of the museum is designed to be a part of the display system, an idealised 'cane thicket' of stainless steel pipes supports the stair treads, display cases and glass parapets.

该科学中心和博物馆位于Rocca di Nazzano的综合性纪念建筑区，由13世纪修建的Savelli 城堡及16、17世纪修建的附属建筑组成。

此建筑位于通往要塞入口的公共广场上，其中有该博物馆的临时展厅和会议室、化石博物馆以及河流博物馆的永久性部分。

此博物馆是一个充满趣味的多样化空间，游客在此可积极地体验复杂多样的台伯河生态系统。

事实上，采用玻璃作为该项目的主要建筑材料唤起了人们对河水透明的色彩和光影的回忆，而这正是该博物馆所致力营造的场景效果。

我们的目的就是尽可能多地缩短游客和展品之间的距离，使展示柜能够引起观赏游客的极大兴趣：这些展示柜并不只是展品的“容器”，它们是人们能够进入的真实“环境”。例如，人们可以聆听河流环境的声音或直接在理想化的河床上行走。

博物馆第一层是由酸蚀玻璃构成的连续“皮肤层”，即含有信息板、透视画、背投和悬臂式展示柜的2.4米高的玻璃墙：游客可以在这本“三维”书上阅读文字、欣赏有关河流生态系统的影像。

在第二层，游客可以进入两个“环境展示柜”内部，从视觉和听觉两方面欣赏河流环境的独特特性。

甚至连接两层之间的楼梯也被设计成为此展示系统的一部分，理想化的“不锈钢管路密集屏障”支撑起了楼梯踏板、展示柜以及玻璃护墙。

LA · BIENNALE

'I International Venice Biennale of Architecture

六届威尼斯国际建筑双年展

udio in the Italian Pavilion | Design: n!studio Susanna Ferrini, Piero Fumo, Paolo Monti, Roberto Rosati, Laura Rossi, Davide Sani, onello Stella | Consultants: Carolina De Camillis, Riccardo Fibbi Lighting Design | Client: Biennale Di Venezia | Contractor: Officine nini Srl Pitigliano (Gr); Vetrerie La Corte Srl Montefiascone (Vt) | Period: 1996 | Model: n!studio

In 1996 we were invited to present our work in the italian pavilion at the vi venice biennale of architecture. The theme was that of the reconstruction of a fragment of a built work. However, there was something not entirely convincing about the idea: what is the sense of rebuilding a fragment of a realty that already exists, without making any further reflection that would place it in a different time and place? We thus decided to create a space where the fragment of reality became an opportunity to reflect upon the deferred perception of architecture: the cantilevered stair of the pitigliano museum was thus recreated on an immaterial wall of glass, exposing the continuous, bent beam that supported it, not visible in the museum, and the photographs and drawings of the completed museum were inserted inside liquid crystal glass panels, rendered transparent by a photo-cell activated by the public. The exhibition thus became an opportunity to reflect upon the various means of perceiving the same reality, of the distance between reality and its perception.

1996年，我们被邀请参加了第六届威尼斯建筑双年展，在意大利国家展团展示我们的作品。展示的主题是对建成建筑的某一部分进行重建。但是，这一概念有一些并不令人完全确信之处：如果不将其设定到一个不同的时间和空间，对现存建筑某一部分进行重建的意义又何在呢？于是我们决定创建一个空间，使现有建筑的某一部分能够体现对未来建筑学的感悟：我们在无形的玻璃墙上重塑了Pitigliano博物馆的悬臂式楼梯，袒露出了其连续、弯曲的支撑梁，而这些支撑梁在该博物馆中是看不到的。我们还将完成后的博物馆的照片和图纸插入到了液晶玻璃嵌板中，通过共同激活的光电管可使其保持透明。这样此展品就能够反映出认知同一实体的不同方式，反映出现实和感悟之间的距离。

Ugo Mulas Exhibition

Ugo Mulas展览

Maxxi, Rome | Design: n!studio Susanna Ferrini, Antonello Stella | Team: Stefano Battaglia, Simone Bove, Paolo Bellisario , Francesca Fontana
Consultants: Carolina De Camillis, Riccardo Fibbi Lighting Design | Client: Maxxi – Museo Nazionale delle Arti del XXI Secolo | Period: 2007

Roberto Crippa

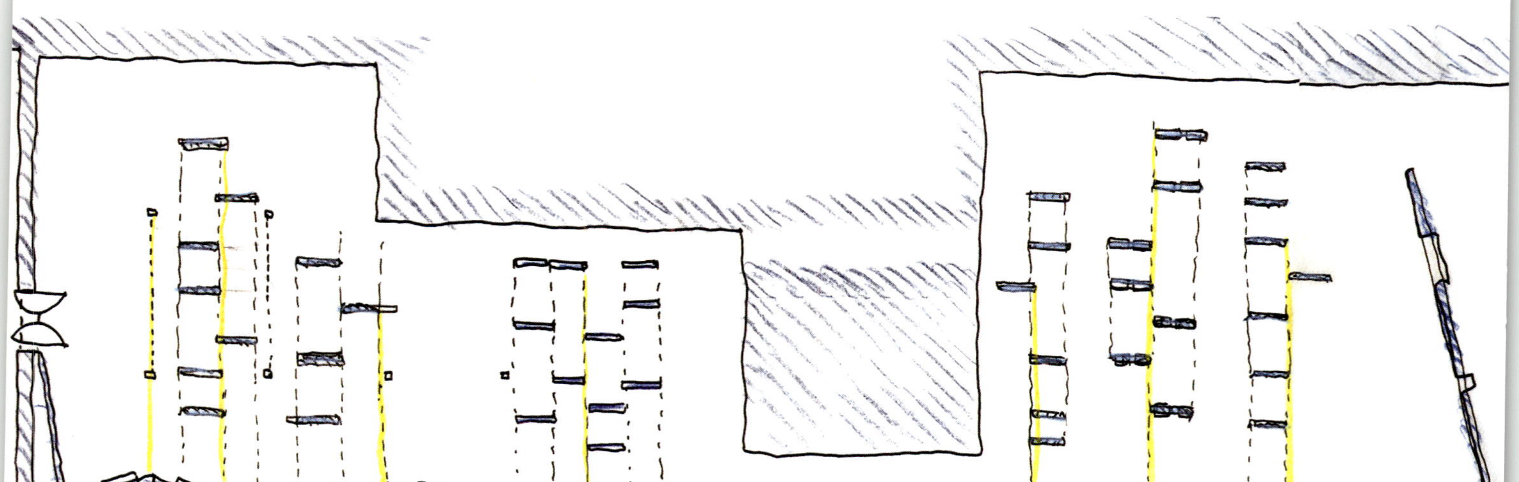

Carla Accardi
Carla Accardi

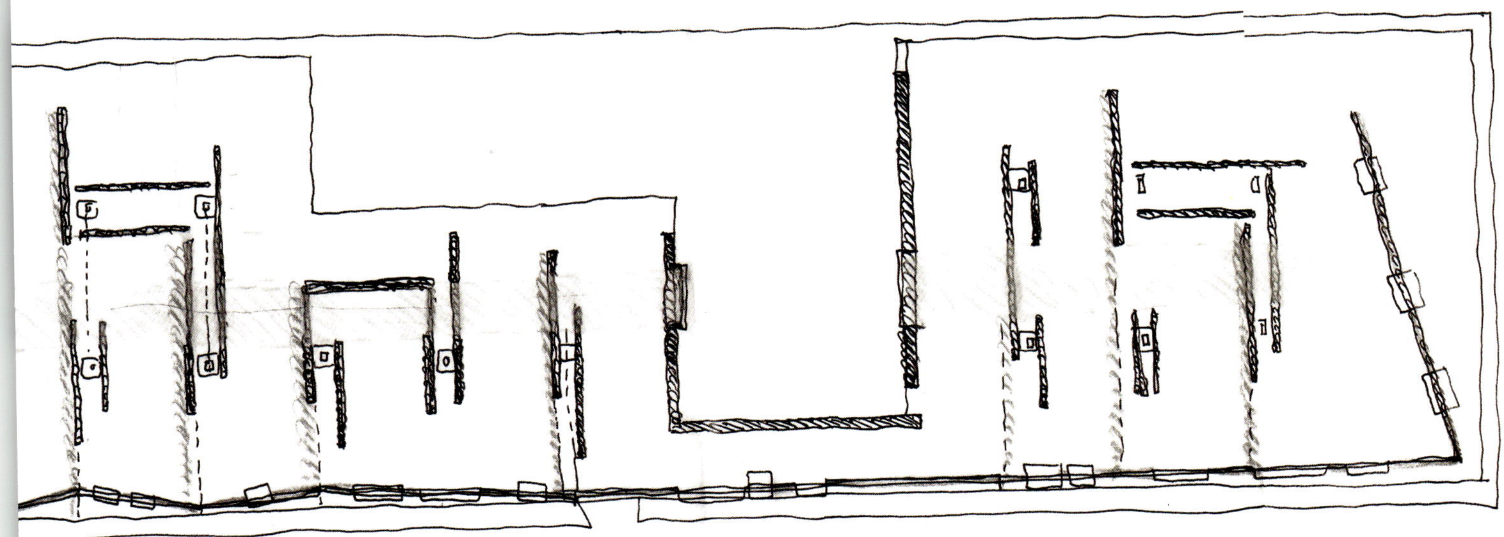

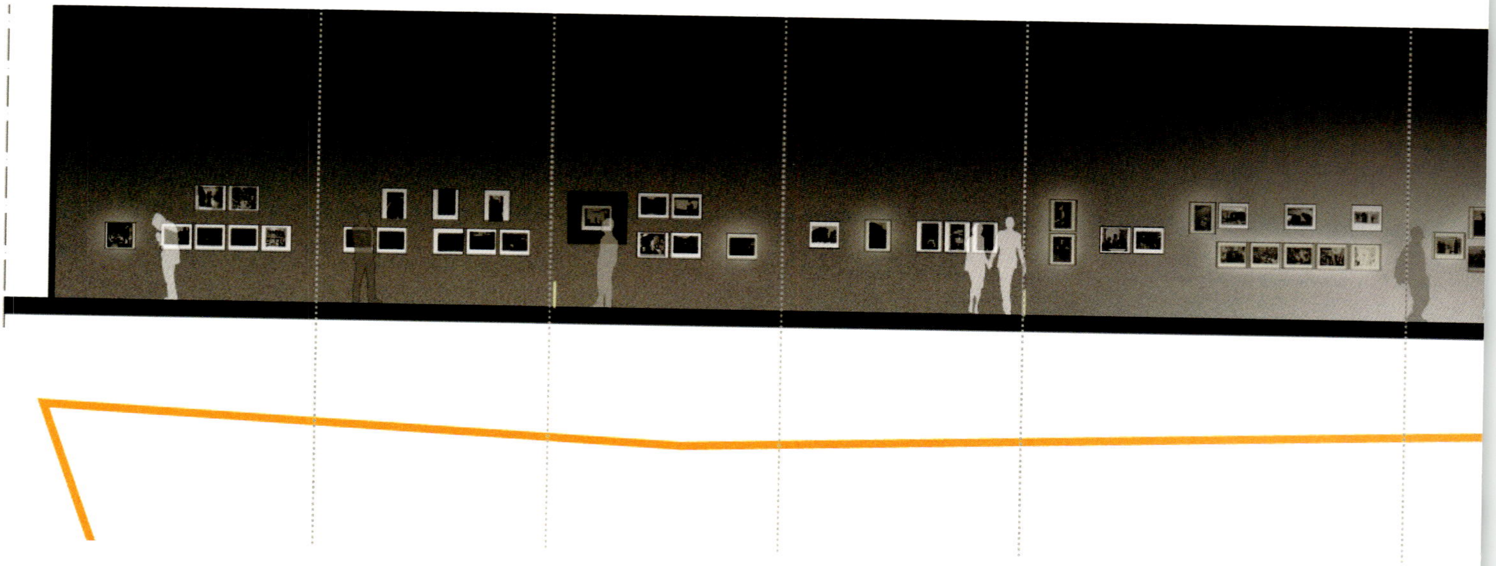

Didascalia

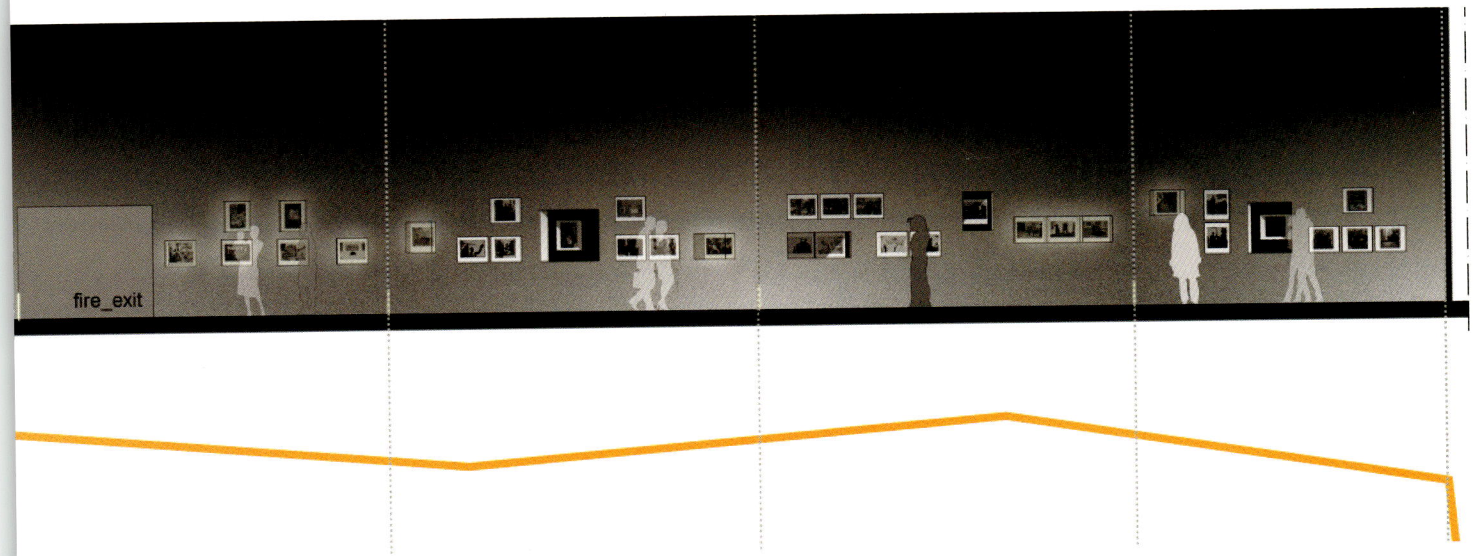
fire_exit

This large exhibition – over 300 photographs capturing the 20 most important years of Italian art, divided into three sections and displayed in two rooms – was presented in the spaces of the former barracks used to house the MAXXI museum as it awaits the completion of its definitive new home designed by Zaha Hadid. The exhibition space, composed of two rooms separated by a central block of services, was defined by a large sidewall that connected the two spaces, and by two central rows of cast iron columns. The dense presentation of the Venice Art Biennale (from 1954 to 1972) was presented on the long wall in a sinuous chronological order, marking and structuring the visitor's path. This led to the open and fluid space of the first section of artists' portraits and the second section of performances-events of experimental art, presented in a system of open boxes made from suspended panels and opaque, evanescent methacrylate rods that wrapped the cast iron columns and lessened their visual impact. Structure and rhythm were entrusted to the lines of fibre optic lighting integrated into the floor and walls of the biennale section, creating illuminated lines that assisted visitors in orienting themselves, and elements of relation between the large chronological room and the archipelago of panels and events in the centre of the display halls.

此大型展览展出300多张照片中记录了意大利艺术最重要的20年，分为三部分在两个展室中展出。在由Zaha Hadid设计的新博物馆完工之前，此展览在曾用作兵营的MAXXI博物馆展出。第一部分展区由一个中间服务区分隔开来的两个展室构成，连接两个展室的大型侧墙由一排铁铸的柱栏构成。在长墙上按照年代的顺序展出了曾在威尼斯艺术双年展（1954~1972年）上展出的作品。而后便过渡到了展出第一部分展区艺术家的肖像和第二部分实验艺术的表演事件回顾的开放性流动空间，展品分别在一种由悬浮板和外面包有轻盈的不透明丙烯酸酯杆的铁铸柱栏的开放型展柜系统中展出，调和了其视觉冲击力，光导纤维照明线路被融入到此展区的地板和墙面之中，其结构和韵律还可帮助游客自己掌握方向，并了解按年代顺序排列的大展室和众多板式结构以及展厅中心的事件展区之间的关系。

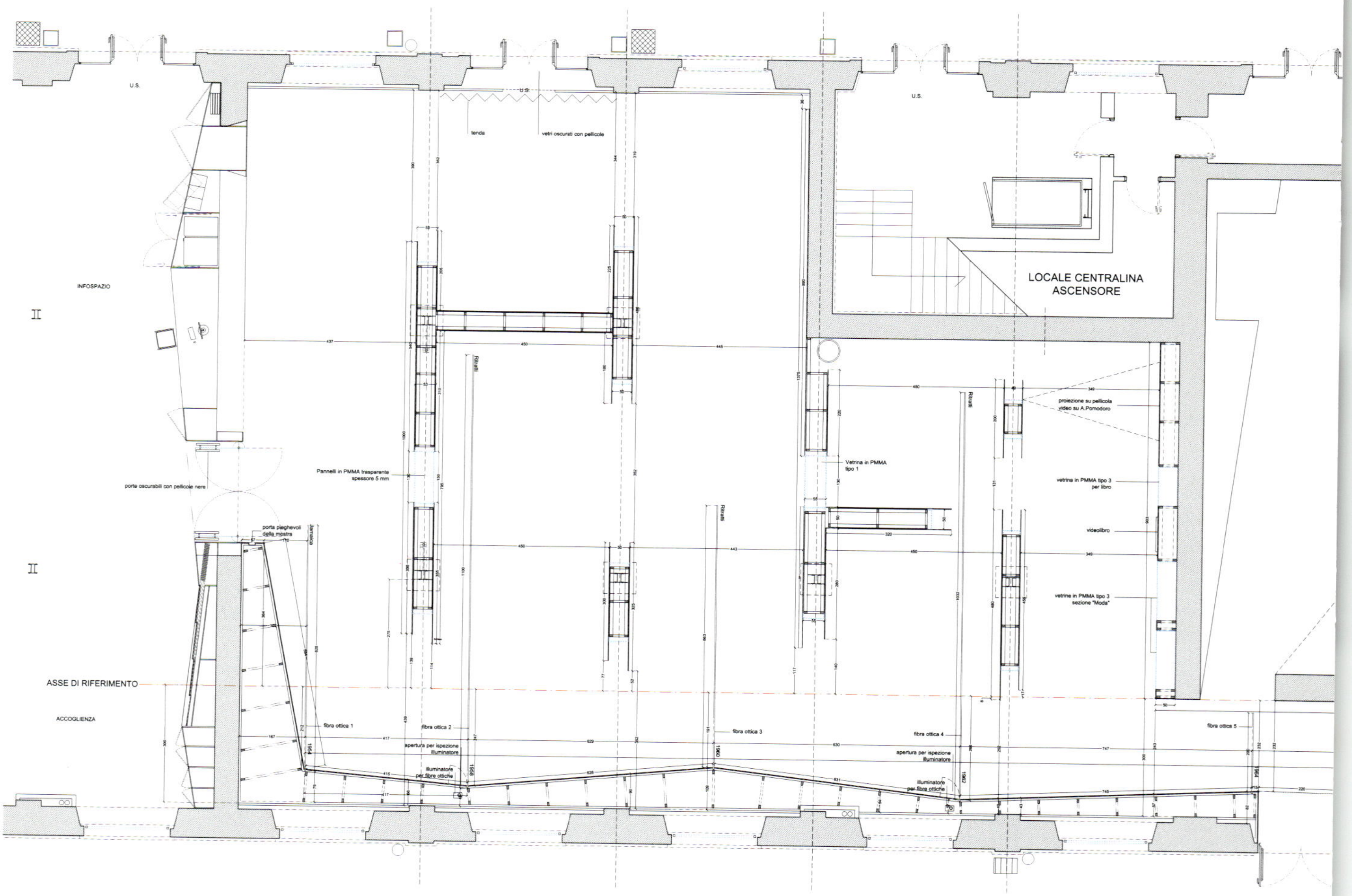

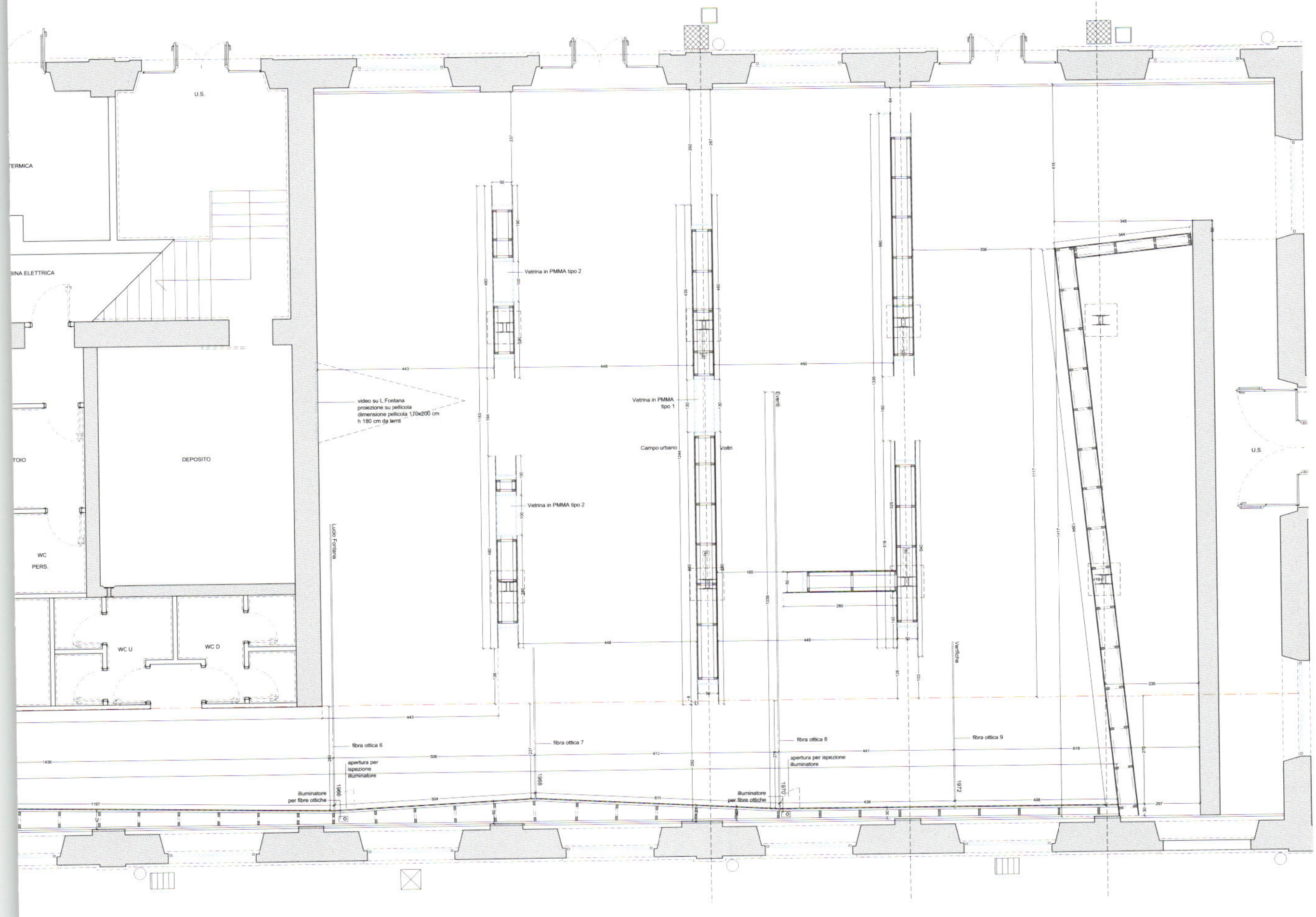
U.S.
DEPOSITO
WC PERS.
WC U
WC D
Vetrina in PMMA tipo 2
Vetrina in PMMA tipo 1
Campo urbano
Voltri
video su L.Fontana
proiezione su pellicola
dimensione pellicola 170x200 cm
h 180 cm da terra
Lucio Fontana
fibra ottica 6
fibra ottica 7
fibra ottica 8
fibra ottica 9
apertura per ispezione illuminatore
illuminatore per fibre ottiche

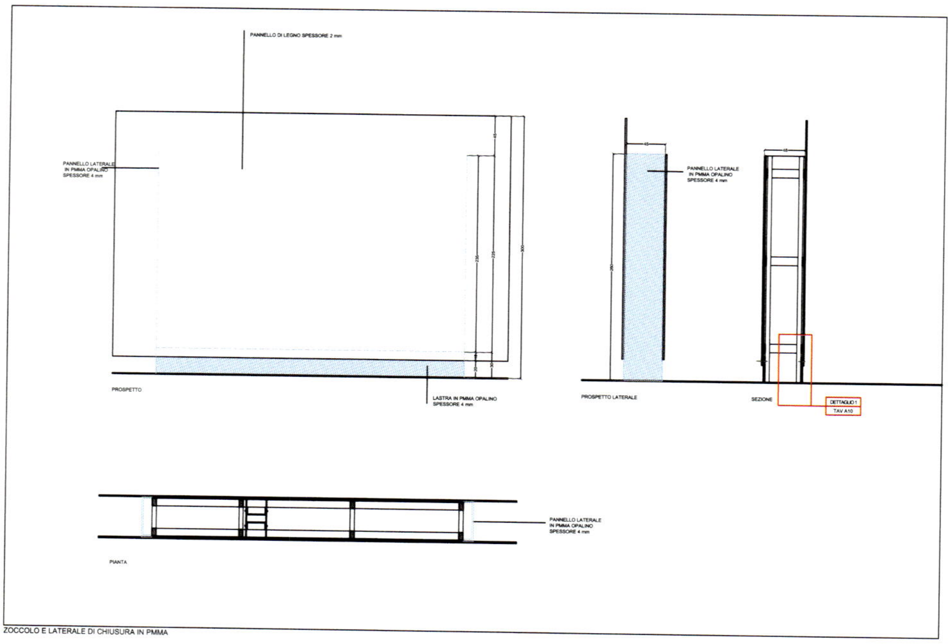

ZOCCOLO E LATERALE DI CHIUSURA IN PMMA

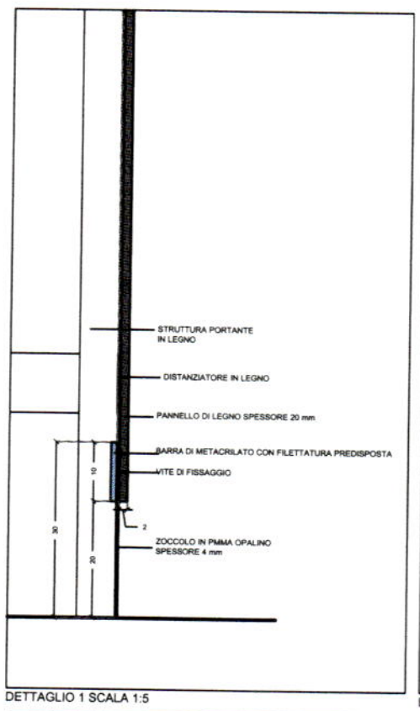

DETTAGLIO 1 SCALA 1:5

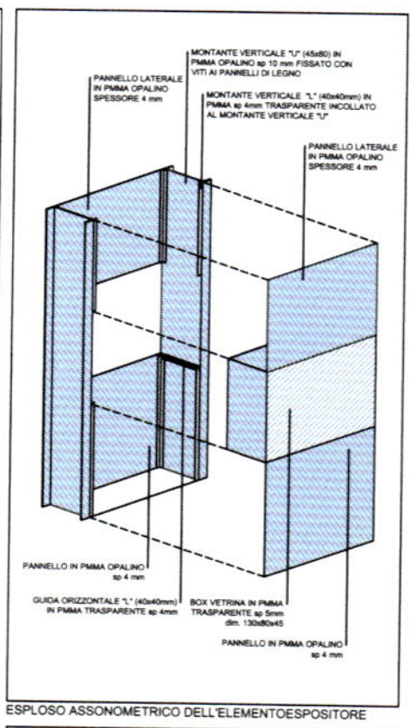

ESPLOSO ASSONOMETRICO DELL'ELEMENTOESPOSITORE

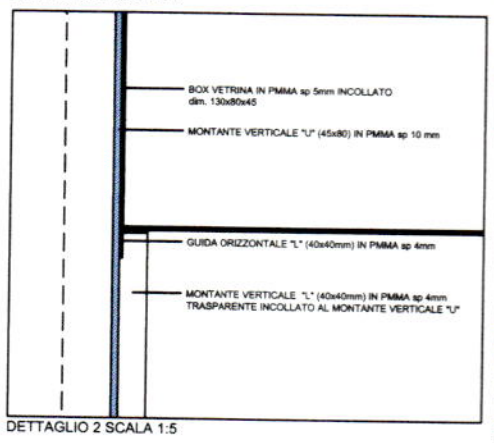

DETTAGLIO 2 SCALA 1:5

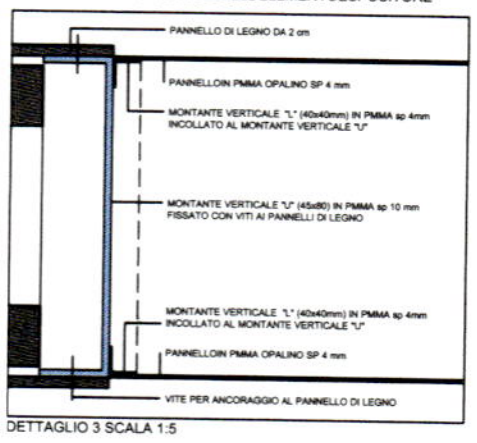

DETTAGLIO 3 SCALA 1:5

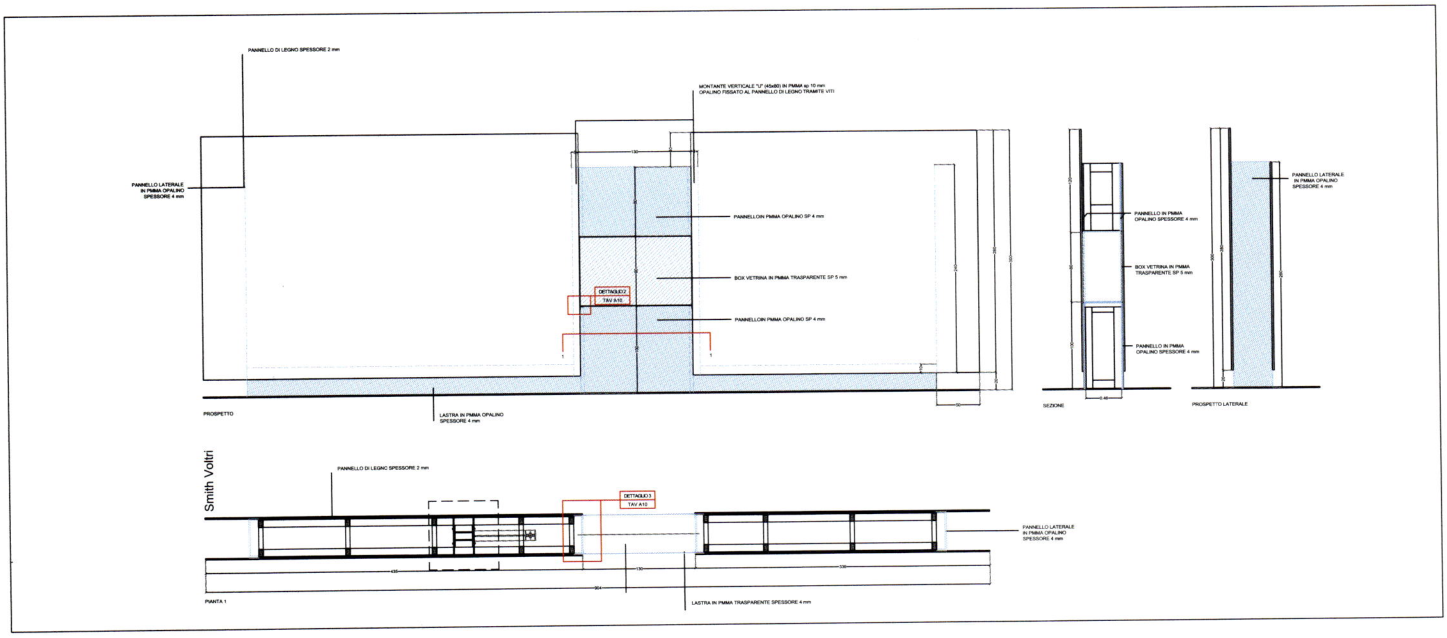
PANNELLO DI LEGNO SPESSORE 2 mm
MONTANTE VERTICALE "U" (45x80) IN PMMA sp 10 mm
OPALINO FISSATO AL PANNELLO DI LEGNO TRAMITE VITI
PANNELLO LATERALE
IN PMMA OPALINO
SPESSORE 4 mm
PANNELLOIN PMMA OPALINO SP 4 mm
BOX VETRINA IN PMMA TRASPARENTE SP 5 mm
DETTAGLIO 2
TAV A10
PANNELLOIN PMMA OPALINO SP 4 mm
PROSPETTO
LASTRA IN PMMA OPALINO
SPESSORE 4 mm
PANNELLO IN PMMA
OPALINO SPESSORE 4 mm
BOX VETRINA IN PMMA
TRASPARENTE SP 5 mm
PANNELLO IN PMMA
OPALINO SPESSORE 4 mm
SEZIONE
PANNELLO LATERALE
IN PMMA OPALINO
SPESSORE 4 mm
PROSPETTO LATERALE
Smith Voltri
PANNELLO DI LEGNO SPESSORE 2 mm
DETTAGLIO 3
TAV A10
PANNELLO LATERALE
IN PMMA OPALINO
SPESSORE 4 mm
PIANTA 1
LASTRA IN PMMA TRASPARENTE SPESSORE 4 mm

Rm008 Exhibition Beijing

北京Rm008展会

Design of the RM_008 Italian Architectural Office Exhibition | Design: n!studio Susanna Ferrini, Antonello Stella | Team: Simone Bove, Sara Ferrazzoli, Paolo Rossi | Consultants: Carolina De Camillis, Riccardo Fibbi Lighting Design | Client: Archi 100 Beijing
Period: 2006

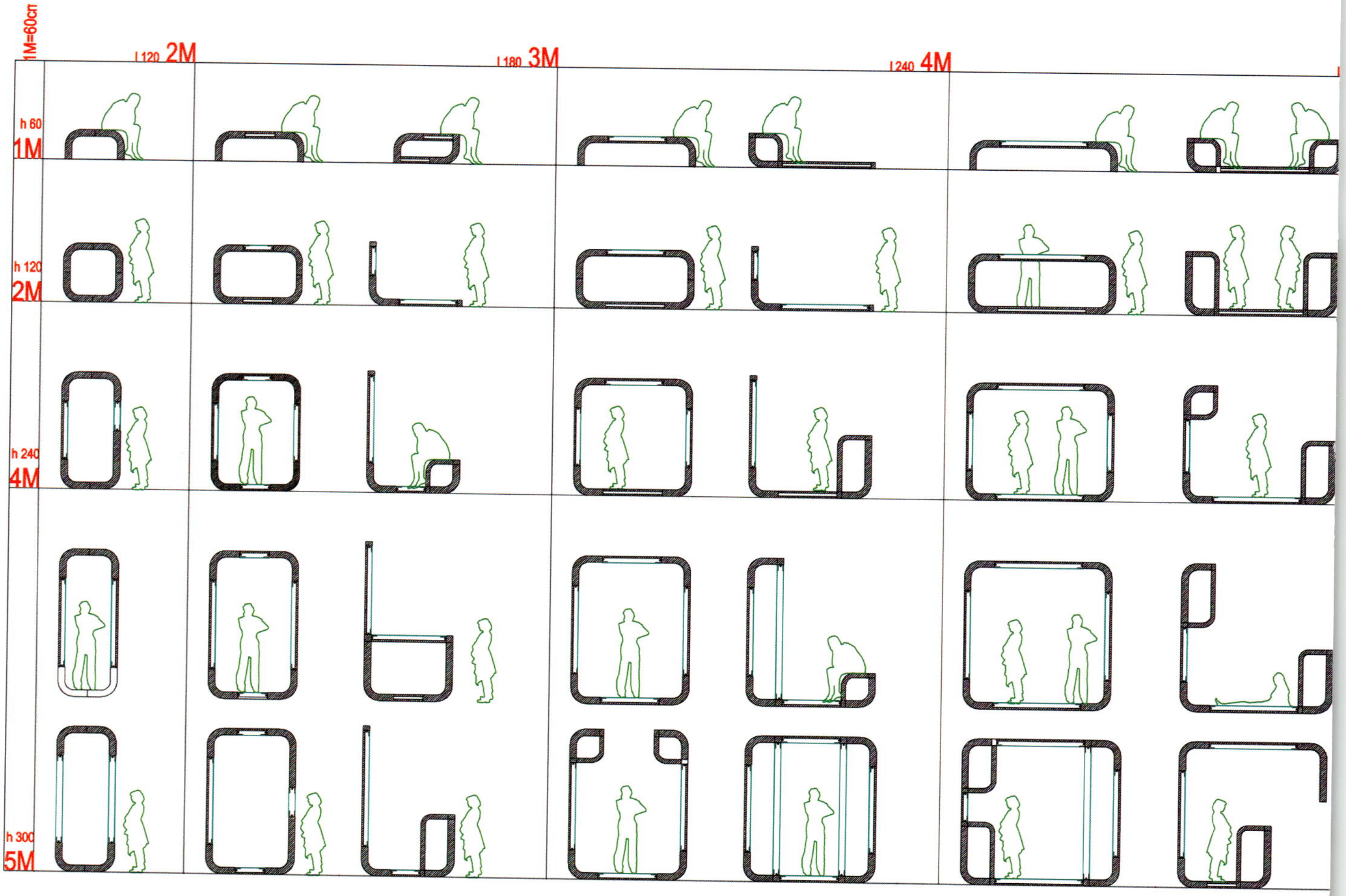
1M=60cm
l 120 2M
l 180 3M
l 240 4M
h 60
1M
h 120
2M
h 240
4M
h 300
5M

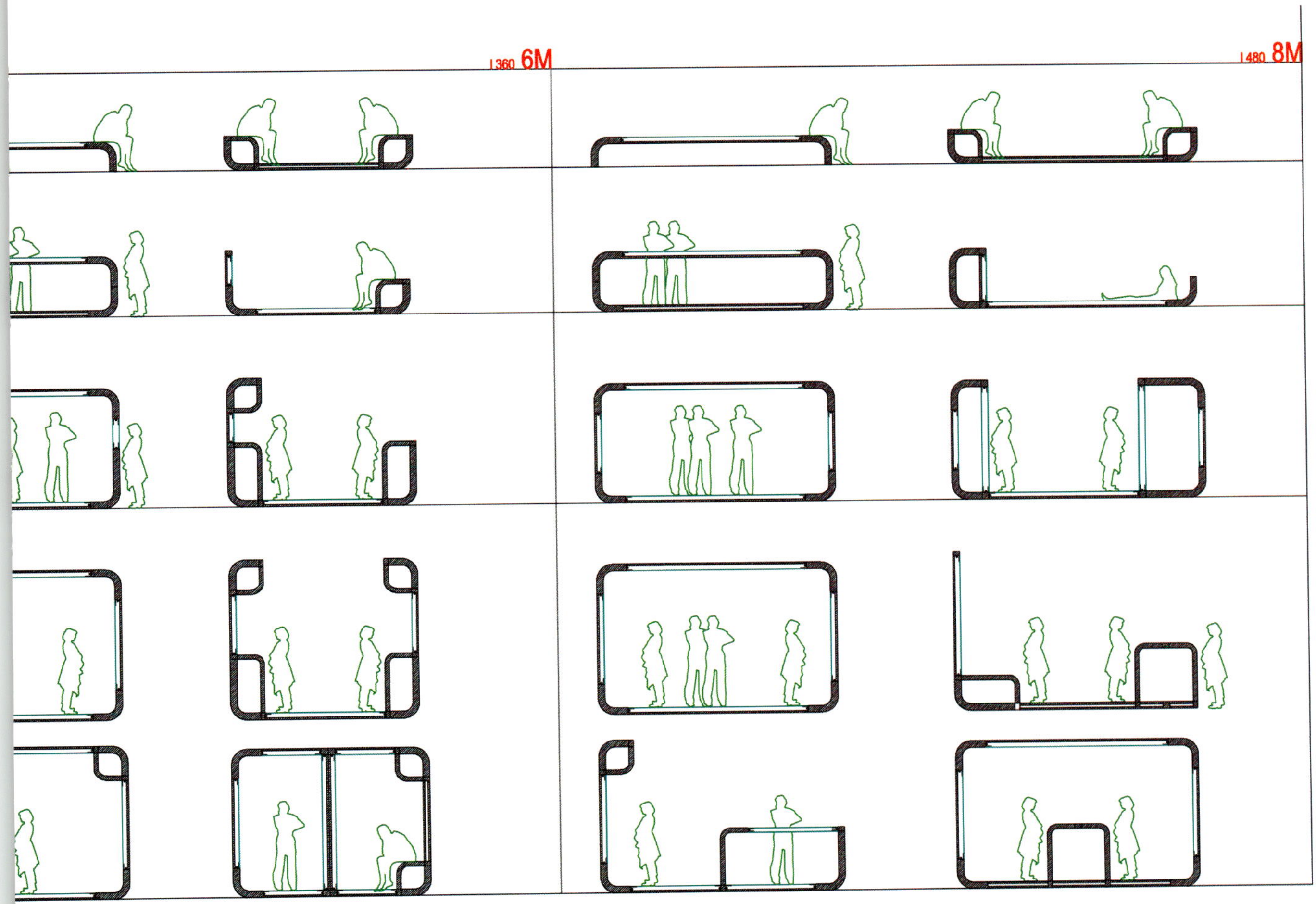

1360 6M
1480 8M

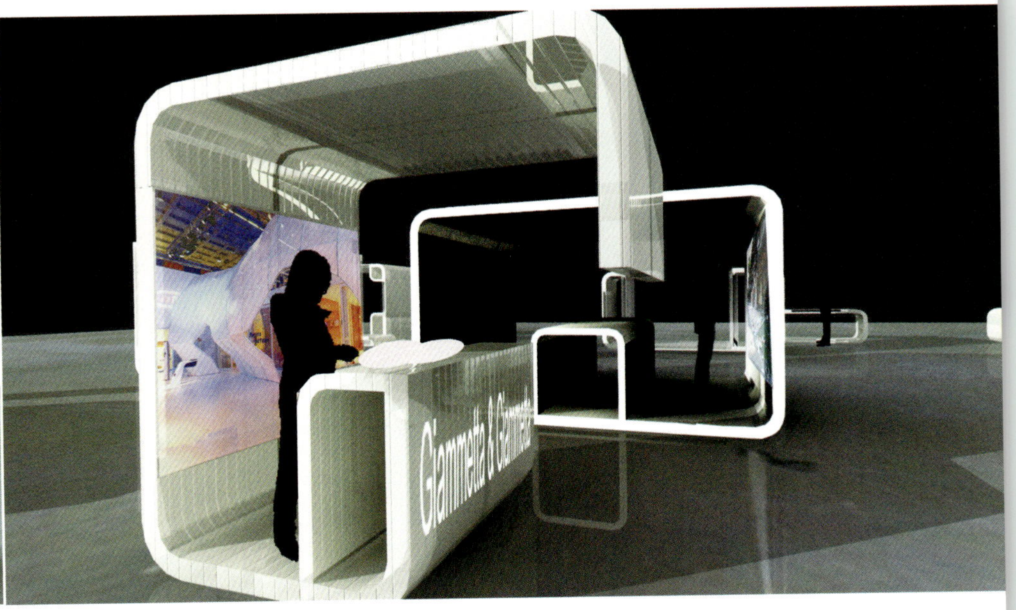

The exhibition is the result of the decision made by a group of nine roman architectural offices rm_008 to present their work during an exhibition proposed by the istituto per gli scambi internazionali archi100 of beijing; an exploration of italian cities and a description of the italian landscape. The display route is developed around the idea of rooms-display containers that can assume different configurations in space. The interior of the rooms simultaneously engage three spheres: the physical perception of the visitor, through the choice of the body's position, induced by the arrangement of the room; visual, audio and tactile perception, through images, projections and multimedia effects; graphics, through which the visitor assimilates information and is guided through the exhibition. The construction system is designed as an assembly of modular elements, comparable to an abacus, which genetically contains the possible and diverse spatial configurations. The primary structure of the room is composed of angular elements and rigid panels of polycarbonate, alternating between transparent and opaque, with built-in video systems and supports for the display of the exhibition contents.

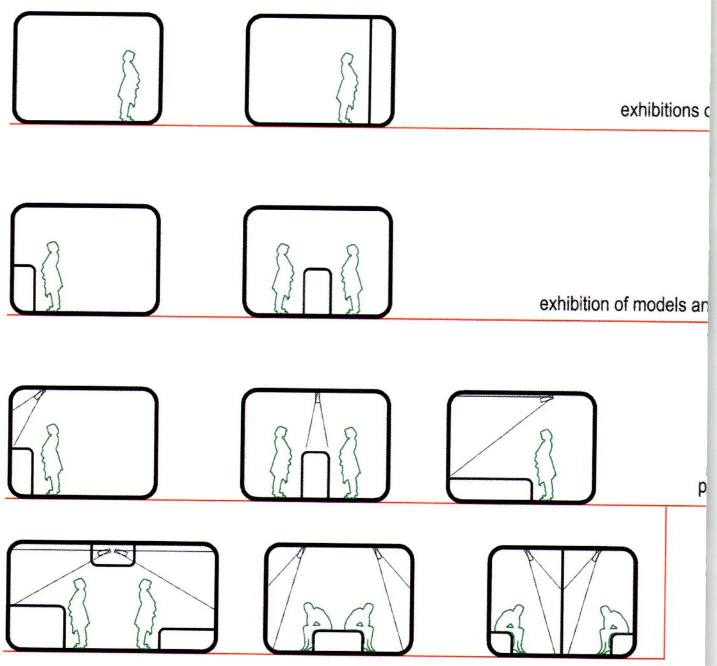

举办此次展会是由9家参与罗马RM_008的建筑事务所共同做出的决定，其目的是在一项由北京Istituto per gli scambi internazionali Archi100邀请举办的展会上展示其作品，探索意大利的城市并介绍意大利的风景。其展示路径围绕着可以采取不同空间配置模式的房屋型展示厅开发布置。这些房屋的内部同时具有三个领域层次：游客可以通过根据房屋中的安排模式选择不同的方位；通过其身体感知展品；通过影像、投影和多媒体效果从视觉、听觉和触觉的角度来感知展品；或通过阅读图表吸收信息并接受游览指导。其建筑系统设计采用模块化组件进行组装，同算盘一样，其本身即具有多种可能和变化多样的立体构型。这些房屋的主要结构由聚碳酸酯角料和刚性板构成，透明带和不透明带交替，并配备有内置型视频系统，可以演示所展览的内容。

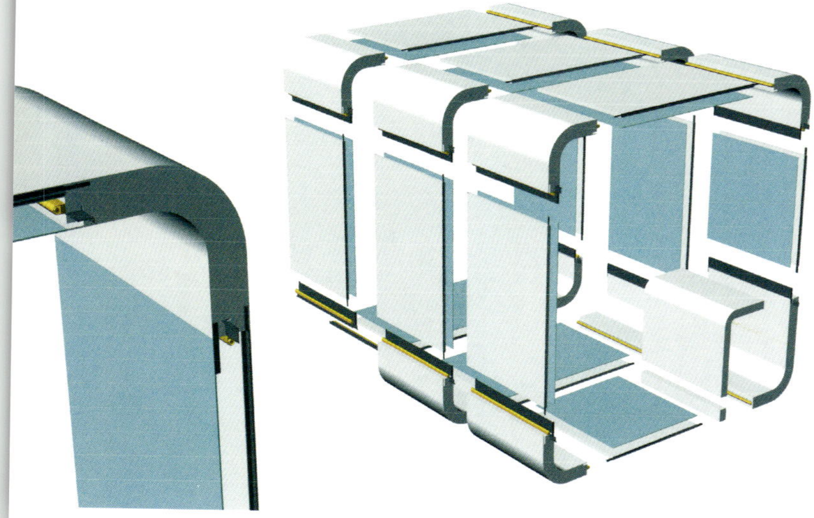

layout of possible planimetric distrib

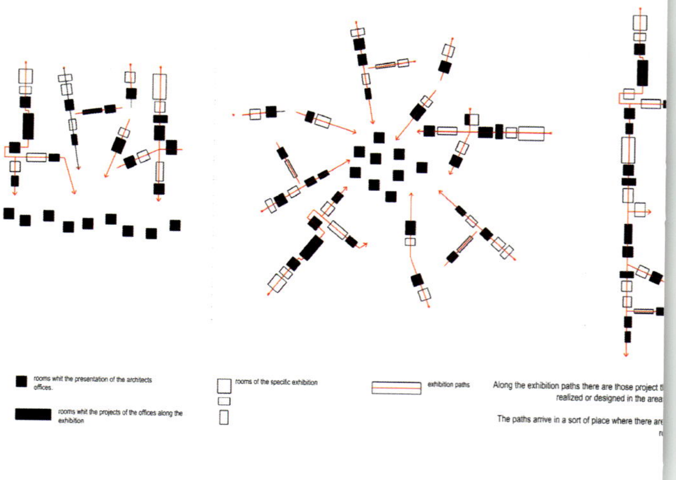

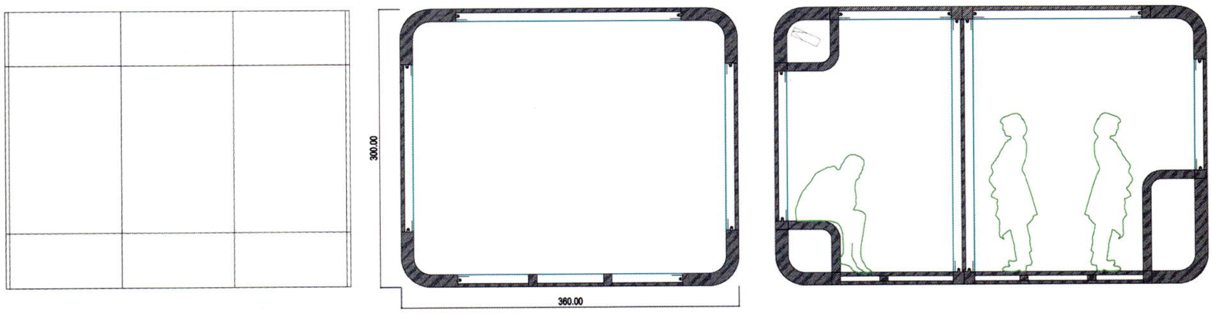

The constuctive system is thought as a mountable system of modular elements, of different shapes and dimensions. In this way it is posible to answer to each exhibition requirement.

The primary stucture of the rooms is made of angular elements and rigid pannels of extruded polycarbonate. The interior of the room is furred with panel of polycarbonate in sheets and a secondary thiner structure of extruded polycarbinate allow to give the rooms differents configurations.

The polycarbonate elements are fixed by metal joints integrated in the structure.

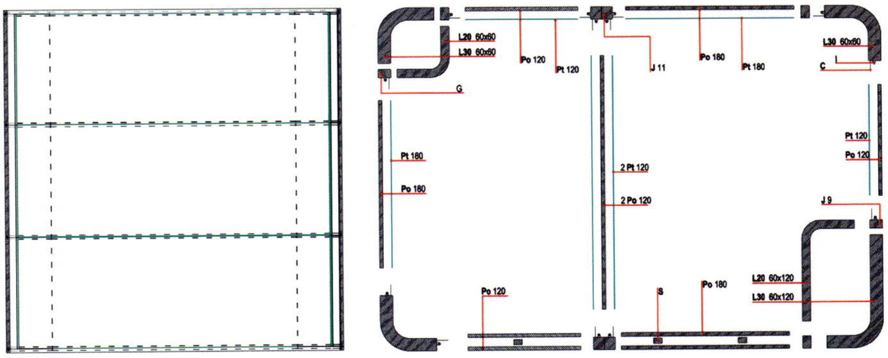

abacus of the costructive elements

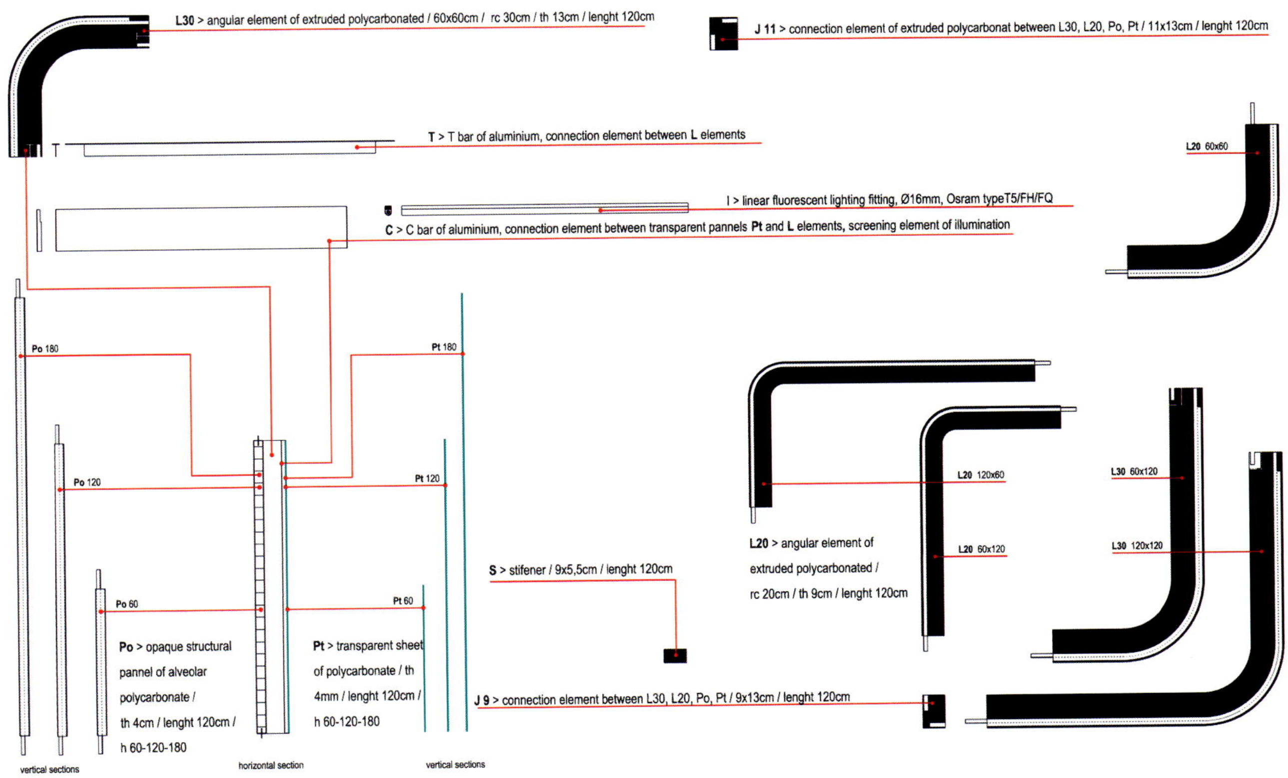

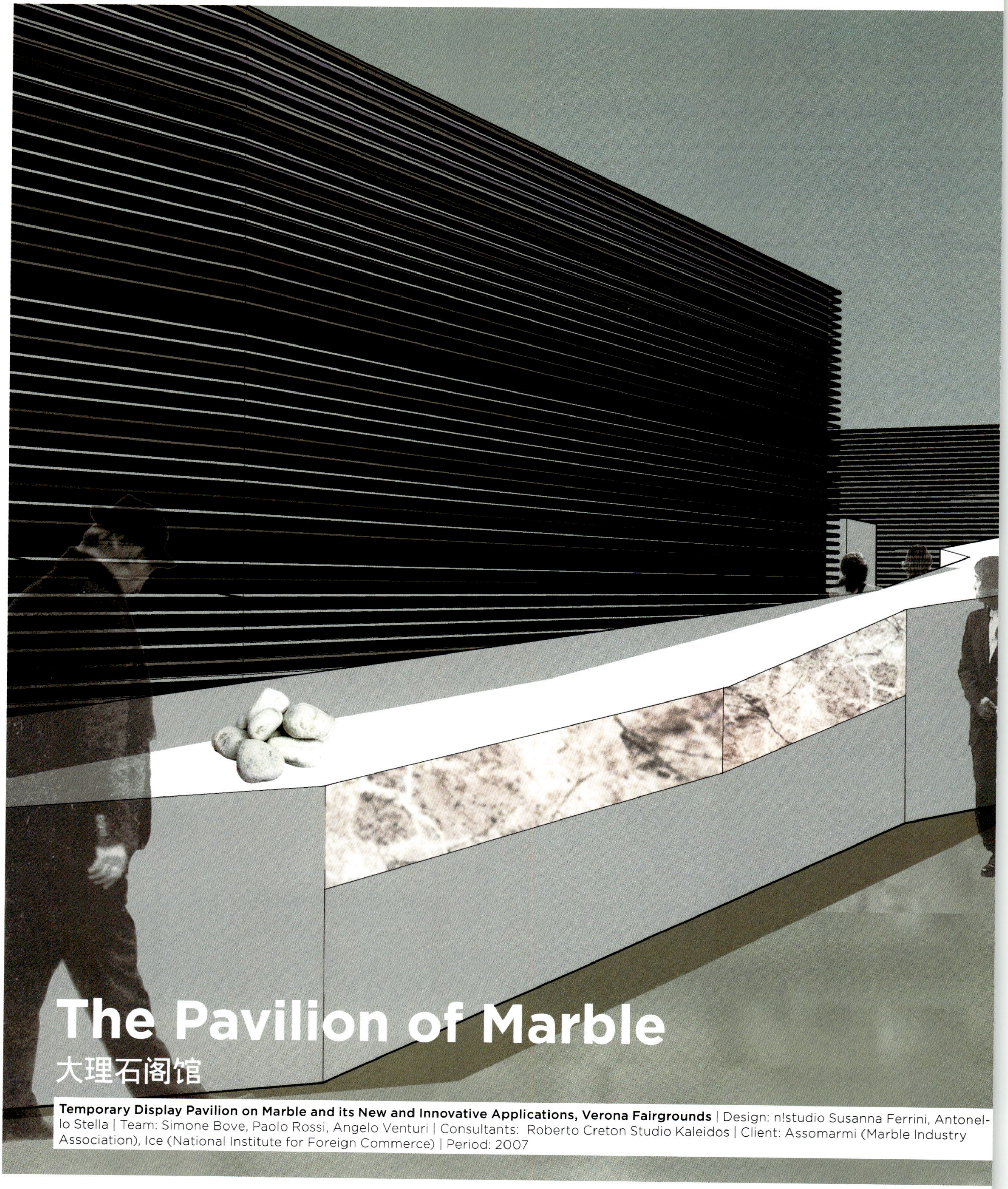

The Pavilion of Marble

大理石阁馆

Temporary Display Pavilion on Marble and its New and Innovative Applications, Verona Fairgrounds | Design: n!studio Susanna Ferrini, Antonello Stella | Team: Simone Bove, Paolo Rossi, Angelo Venturi | Consultants: Roberto Creton Studio Kaleidos | Client: Assomarmi (Marble Industry Association), Ice (National Institute for Foreign Commerce) | Period: 2007

exhibition

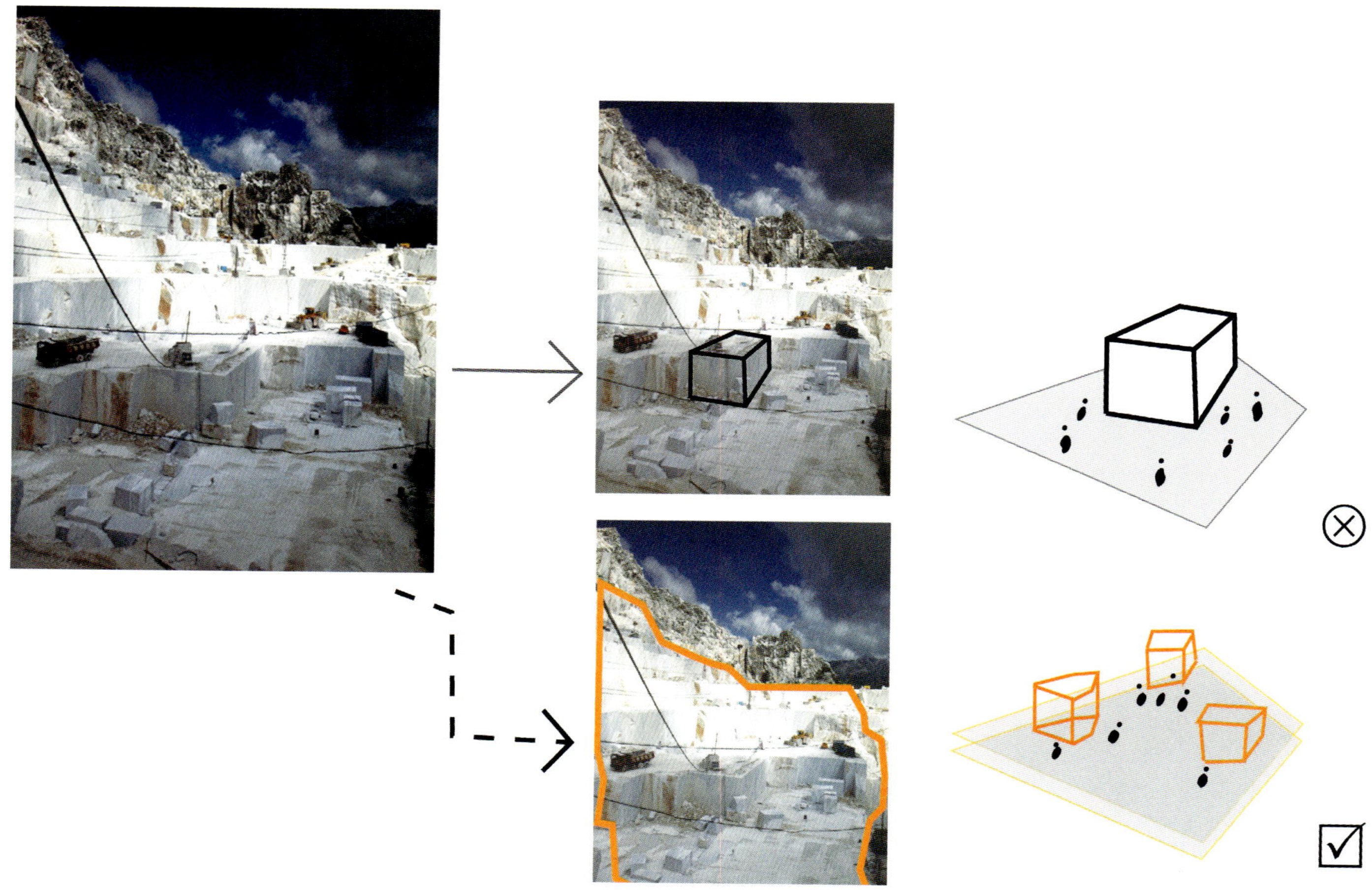

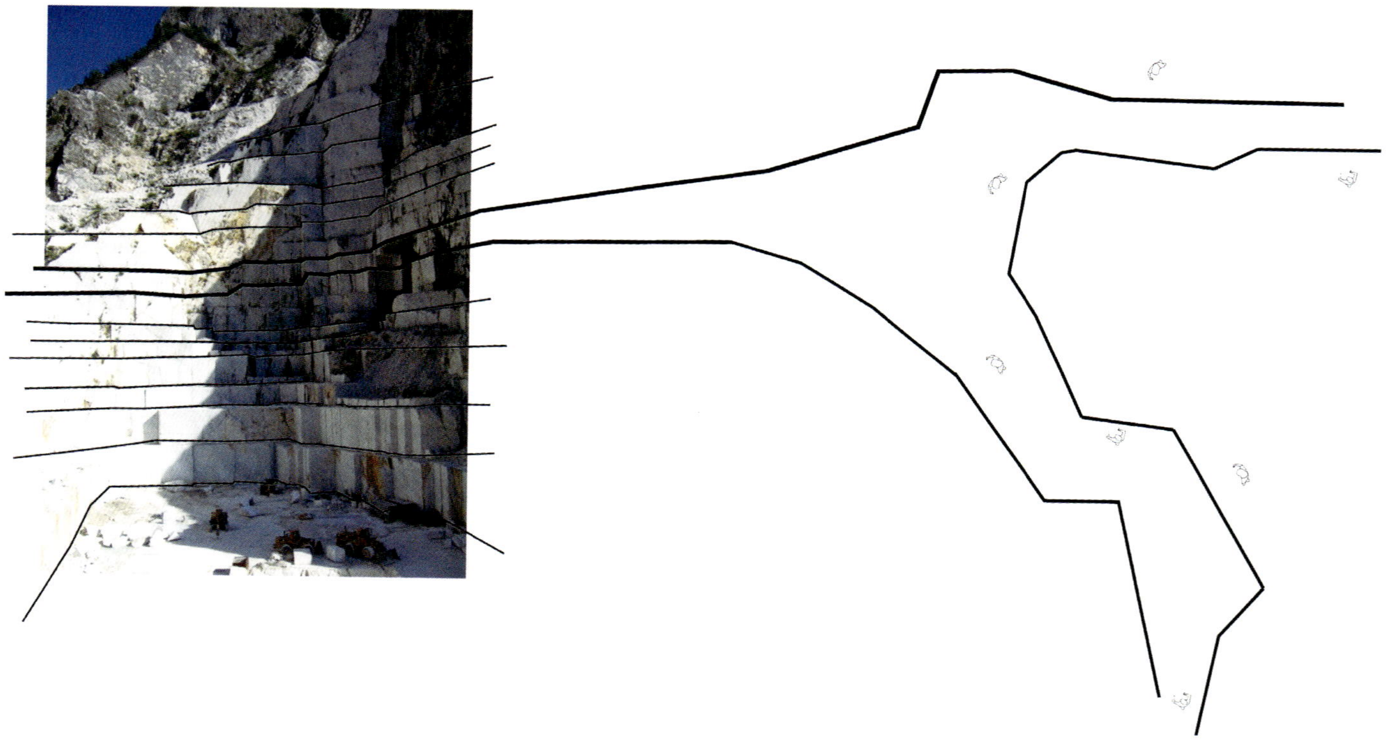

This temporary pavilion was designed for the international stone fair held in Verona (Marmomacc). The project was commissioned by Assomarmi, the Italian association of marble producers and in collaboration with ICE, the National Institute for Foreign Trade.
The request was for an approximately 1,000 m^2 space that would evoke the material in all of its various uses - historical and, above all, innovative - based on the most recent applications of this long-lasting material.
Our starting point for the definition of the space was a reference to stone quarries, a space that ideally unites all marble producers: the concept as developed into the idea of a monumental space rather than a monumental object (typical of specialised fairs). The geometric lines of the quarry are a strong element of the design, and the contour lines generate a space of total immersion.
The pavilion is composed of three complementary elements: the entry block, whose seating is used to welcome and inform visitors and filled with video terminals and information stands; the sensory walk, the heart of the pavilion that cuts into the mass and passes through the three pavilions in a walled and covered space; the display wall, that accompanies visitors towards the exit, in a space whose height varies from 1 to 4 metres, and containing veining-display cases filled with highly technological and innovative marble products.
The three pavilions (4 metres in height) were inserted within the space as a sort of visual rest point (max. 30 visitors at one time), presenting different visions of the world of marble. The first rest area corresponded with the multimedia visualisation of marble, its declinations and/or inspirations, using a back-lit circular projector; the second area corresponded with a tactile path passing through a darkened space in which it was possible to touch and re-discover the various qualities of marble and its multiple finishes; the third space was a mise en scene in a small theatre space where the work presented on stage alternated holographic projections with blocks of marble. Each of the three pavilions was clad with 0.5 mm marble slats that challenged the historical "gravity" of this material, evoking possible new scenarios.

此暂时性展馆是专门为在Verona (Marmomacc)举办的国际石材展设计。此项目由意大利大理石生产厂家联合会Assomarmi同国家外贸研究所ICE联合委托设计的。
设计要求有大约1000 m^{2}的展区，人们能够通过这些经久耐用的材料的展示，了解其各种各样的用途，包括历史应用情况，以及最重要的是其创新应用信息。
作为对此展示区设计规划的起点，我们首先介绍了各家采石场，此展区以理想的模式将所有大理石生产厂家组织到了一起：即开发一个具有重大意义的展区而不是具有重大意义的展品的概念（专业展览会的典型模式）。采石场的几何线条是本设计中的重要元素，这些线条的轮廓形成了一个使观展者完全沉浸其中的展区空间。

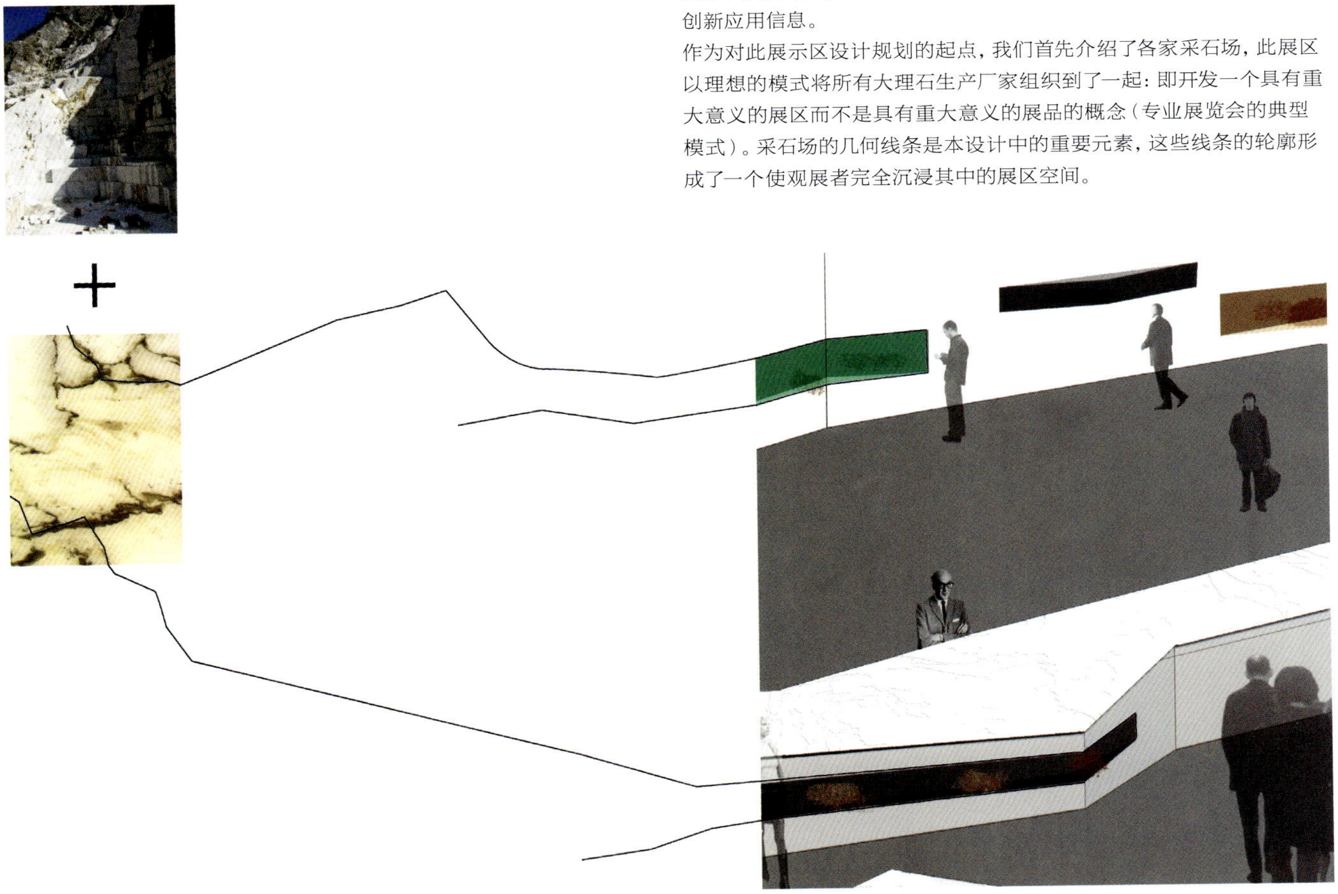

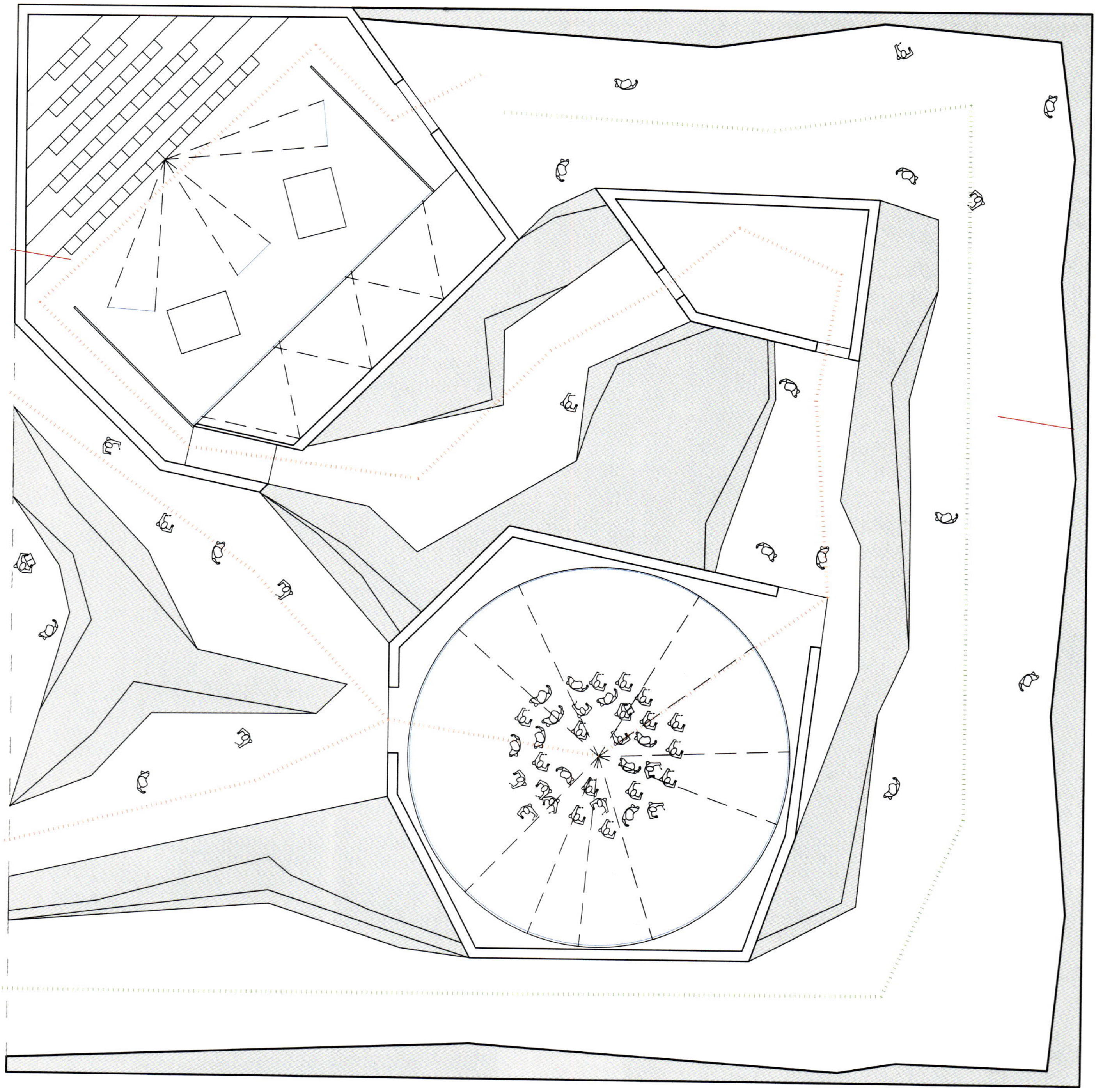

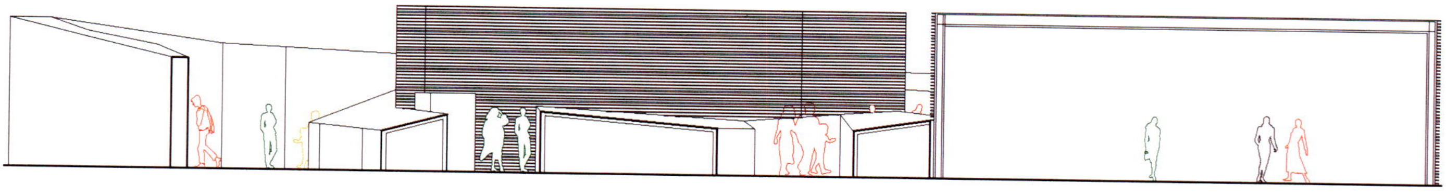

此展区具有三种补充构造：入口区，其底部用于欢迎接待游客之用，充满各种视频终端和信息台；感官走廊，这是此展馆的核心，切入到其主体部分之中，通过具有围墙和顶棚的空间穿过三个展馆；展示墙，一直伴随着游客直到出口之处，其空间高度从1米~4米，呈叶脉状的展示柜用于展示高技术含量且具有高度创新性的大理石产品。
三个展馆（4米高）被插入到整个空间中作为一种视觉休息点（一次最多容纳30名游客），使您能够通过不同的视觉角度欣赏这个大理石的世界。在第一个休息区通过背投式圆形投影仪演示大理石的多媒体视频，展示与其相关的偏差和灵感；第二个区域有一条穿过一个暗室区的触觉通道，在其中游客可以触摸并重新体会大理石的各种质地及其多种抛光性能；第三个休息区是一个具有小型剧场的舞台演出区，以全息投影的方式交替展示了大理石块构成的舞台作品。这三个展馆均覆以0.5mm厚的大理石板条，它是对该种材料有史以来便已存在的“重力”限制的挑战，揭示出了新的可能场景。

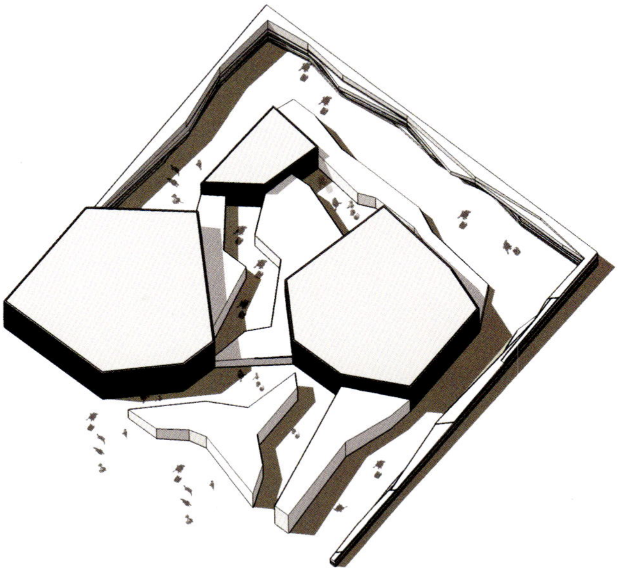

'Mirroring the Age' Beijing

“镜像时代” 北京展

World Art Museum Exhibition Design, Beijing Mirroring the Age 2005 and Great Civilisation 2006 | Design: n!studio Susanna Ferrini, Antonello Stella with Giovanni Bulian (Coordinator) | Team: Simone Bove, Stefano Battagia, Mariangela Corsi, Sara Ferrazzoli, Gabriele Misso, Tiziana Monti, Loris Rossi, Paolo Rossi, Giulia Scaglietta, Teresa Zanaria | Consultants: Carolina De Camillis, Riccardo Fibbi Lighting Design Collaborators: Claudia Campone | Client: Italian Ministry of Cultural Heritage, World Art Museum | Period: 2005 and 2006

The project responds to the requests made by the beijing gehua cultural development group to the create the first world art museum in china, inside the china millennium monument, to promote education about the art of the world's great civilisations.

The first exhibit – 'mirroring the age, men and their world through six centuries of italian art works in the florentine state collections' included a vast selection of masterpieces from florentine museums and was designed to promote cultural exchanges between italy and the people's republic of china during the year of italy in china. The exhibition was designed around the juxtaposed and 'dialoguing' placement of two display walls along the main circulation route, generating a crescendo of dynamism and tension: a glowing pvc wall presented the historic and geographic context of the works displayed in silk screenings and video projections; the other, composed of stainless steel mesh panels, contained display niches and glass display cases, pure transparent and opaque prisms. The exhibition was used to experiment with the flexibility and modularity of assembly, which is capable of adapting itself to the various exhibitions planned by the museum for the coming years.

The successive test was that of the exhibition entitled 'great civilisations', inaugurated in 2006 and using the same display system, arranged in a different layout to meet the theme of this new exhibition.

该项目是应北京歌华文化发展集团的要求而进行设计的，其目的是要在中华世纪坛内建造中国第一座世界艺术博物馆，以促进对于伟大的世界艺术文明的教育。在其第一个展会——“镜像时代”，透过佛罗伦萨的艺术典藏了解6个世纪中的意大利人和他们的生活，展示了大量佛罗伦萨博物馆的艺术杰作，其设计目的旨在促进中国意大利年中意大利和中华人民国和国之间的文化交流。该展会的设计围绕着环形展示路线上的两个并排的展示墙，以一种“对话式”的排放展开，营造出了一个动力和张力的高潮。采用一座PVC发光墙通过丝网屏和视频投影呈现了所展示作品的历史和地理背景信息。而另一座由不锈钢网板构成的墙中则有着展示壁龛、玻璃展示盒、完全透明和不透明的棱镜。该展会用于测试其组装的适应性和模块性，测试其是否能够适应计划将于未来几年中举办的各种展会。随后受到的一次检验是名为“伟大文明”的展会，该项展会于2006年开幕，采用相同的展示系统并重新进行了布局以满足这一新展会的主题要求。

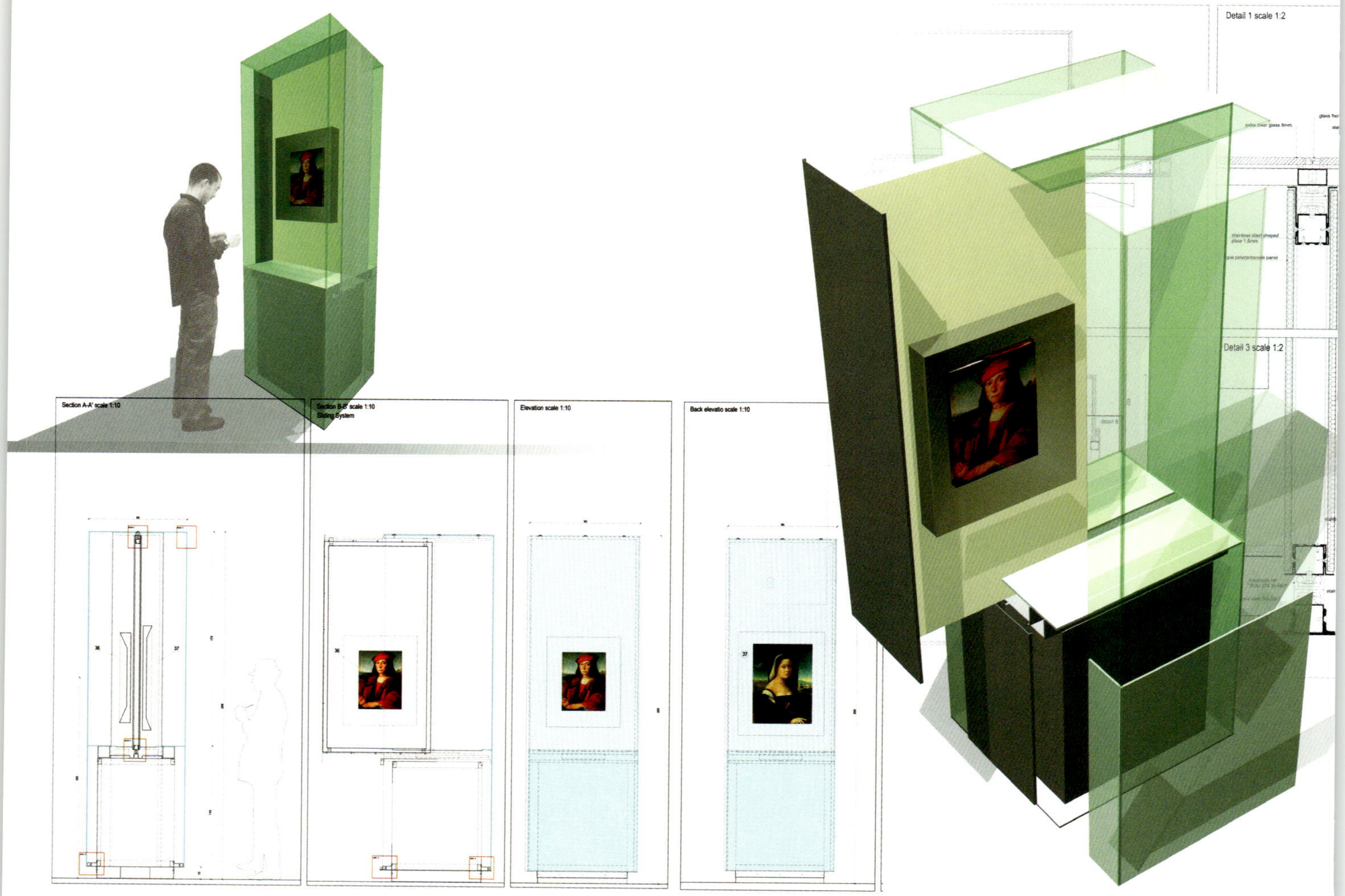
Section A-A' scale 1:10
Section B-B' scale 1:10
Sliding System
Elevation scale 1:10
Back elevatio scale 1:10
Detail 1 scale 1:2
Detail 3 scale 1:2

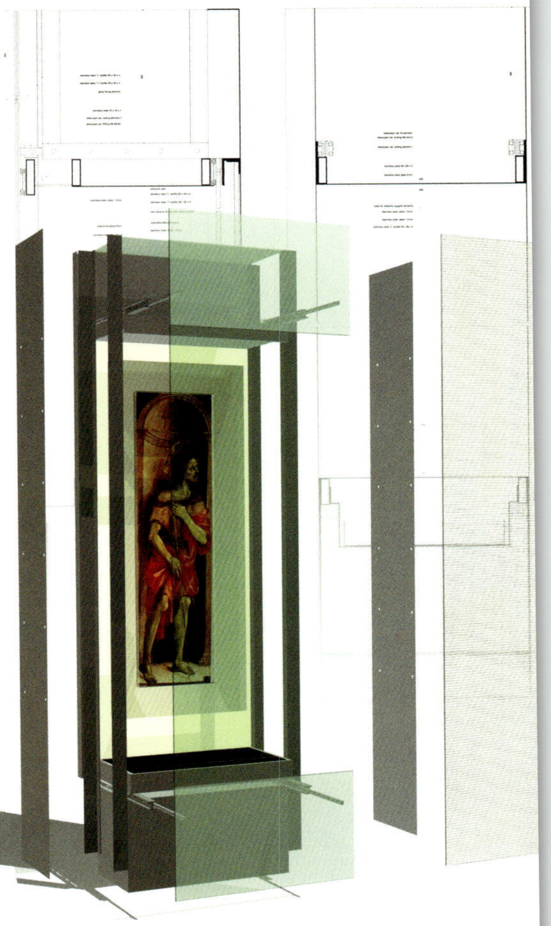

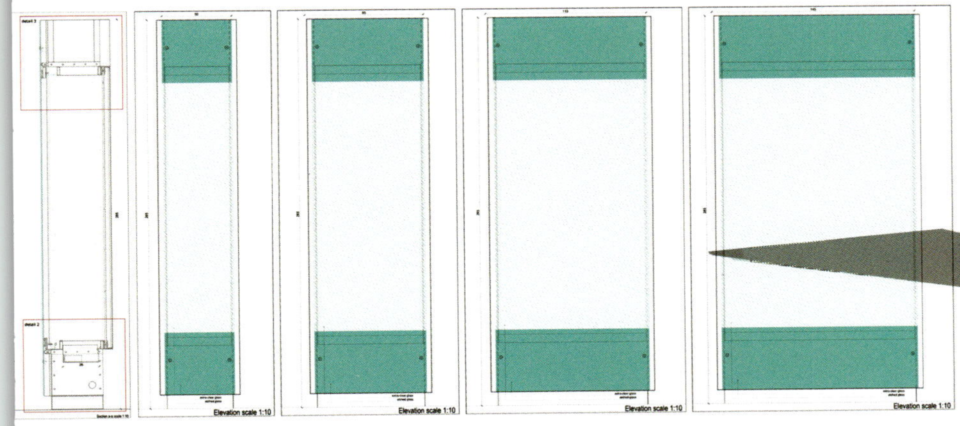
Elevation scale 1:10
Elevation scale 1:10
Elevation scale 1:10
Elevation scale 1:10

Western Market Museum

唐朝西市博物馆

The Tang Dynasty West Market Museum, Xi'an China | Design: n!studio Susanna Ferrini, Antonello Stella with Giovanni Bulian
Team: Simone Bove, Stefano Battaglia, Stefano Bulian, Anna Hollstein, Paolo Rossi, Lara Turchini | Consultants: Carolina De Camillis, Riccardo Fibbi Lighting Design, Dott. Roberto Ciarla Scientific Curator, Wmedia Multimedia | Client: The Xi'an Tang Dynasty West Market Real Estate Co., Ltd | Period: 2008

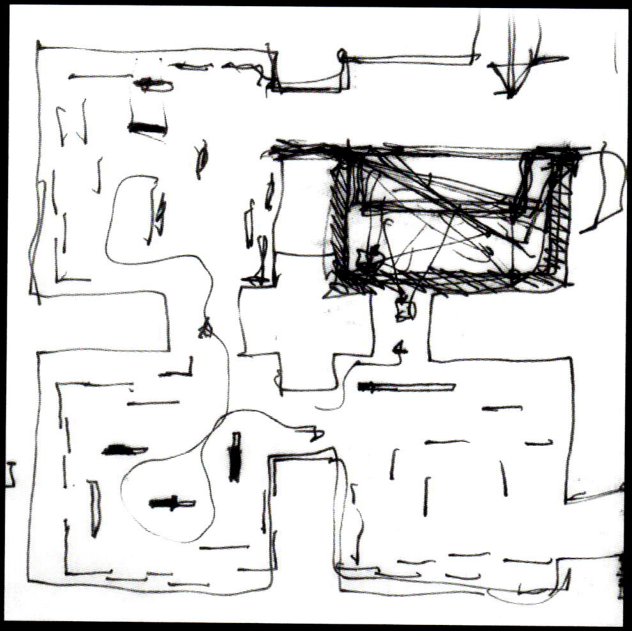

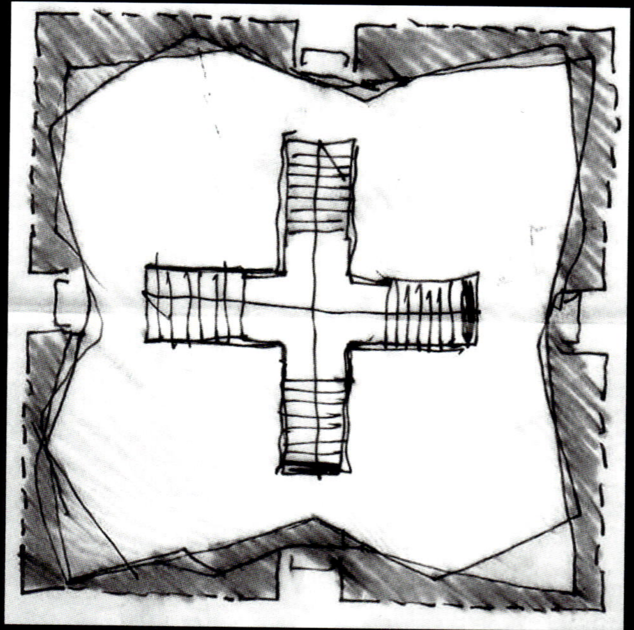

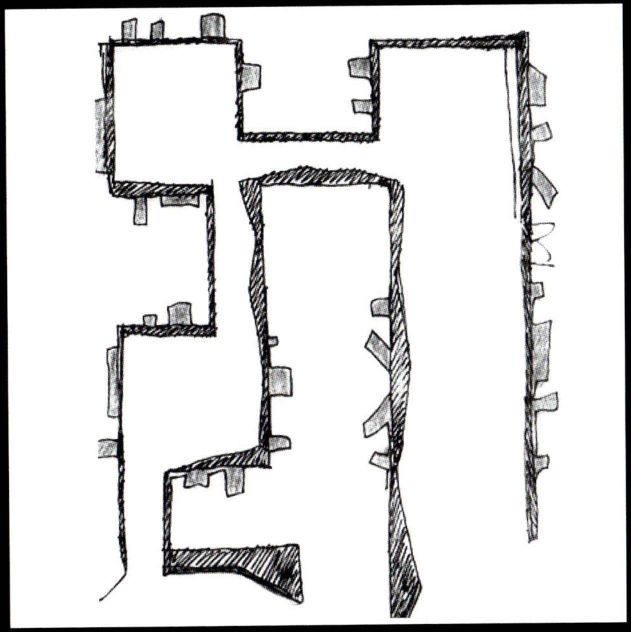

exhibition

FIRST FLOOR 一楼

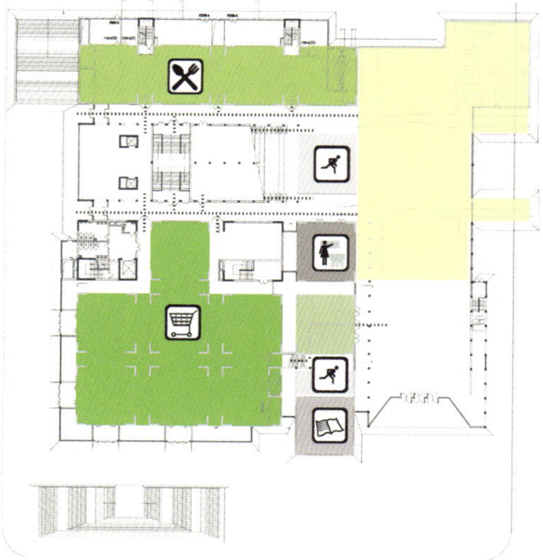

THIRD FLOOR 三楼

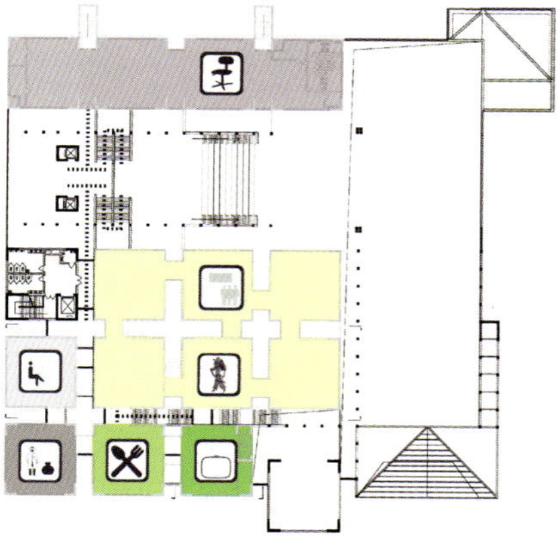

SECOND FLOOR 二楼

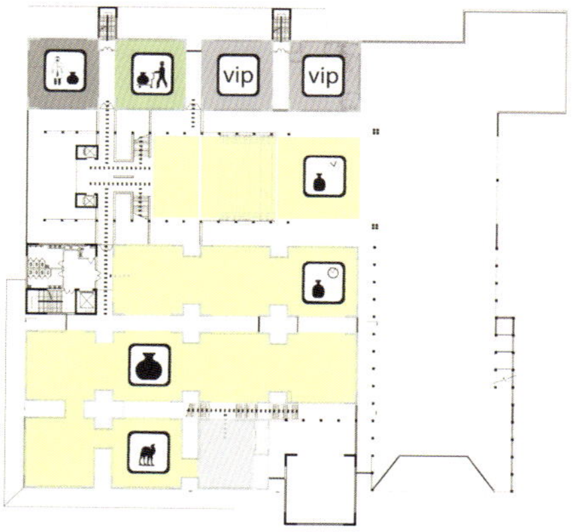

FOURTH FLOOR 四楼

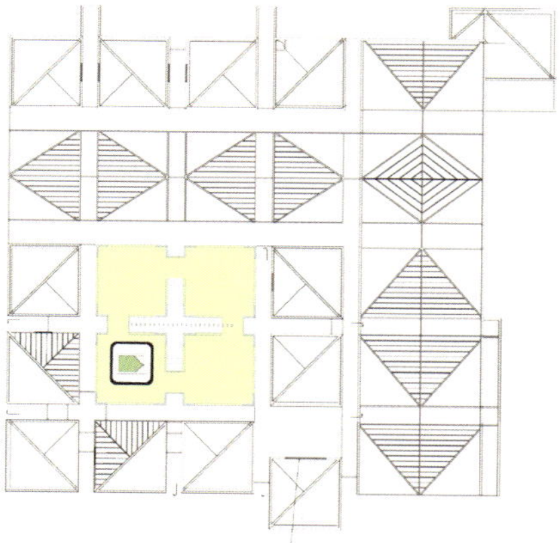

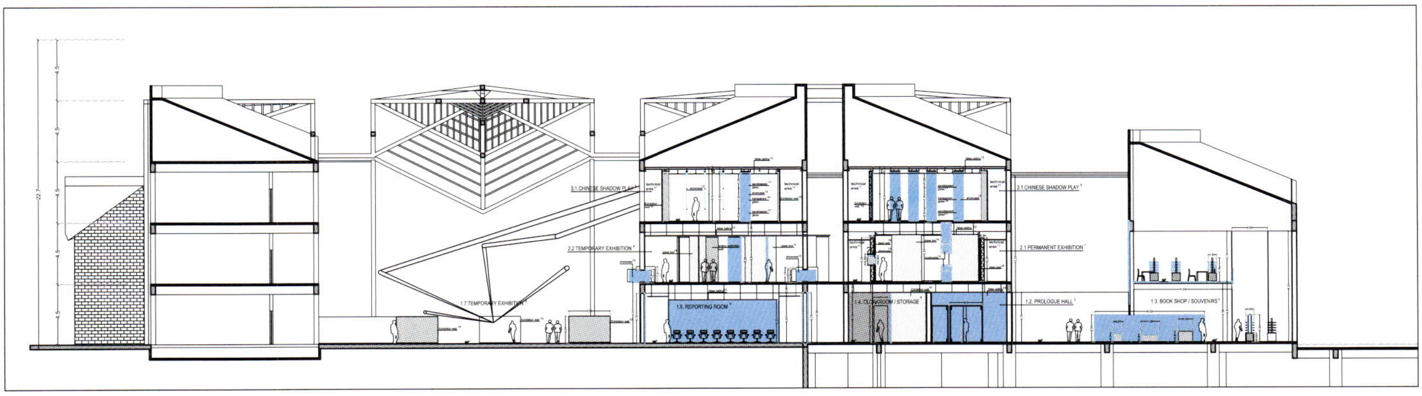
3.1 CHINESE SHADOW PLAY
2.2 TEMPORARY EXHIBITION
1.7 TEMPORARY EXHIBITION
1.8. REPORTING ROOM
1.4. CLOAKROOM / STORAGE
3.1 CHINESE SHADOW PLAY
2.1 PERMANENT EXHIBITION
1.2. PROLOGUE HALL
1.3. BOOK SHOP / SOUVENIRS

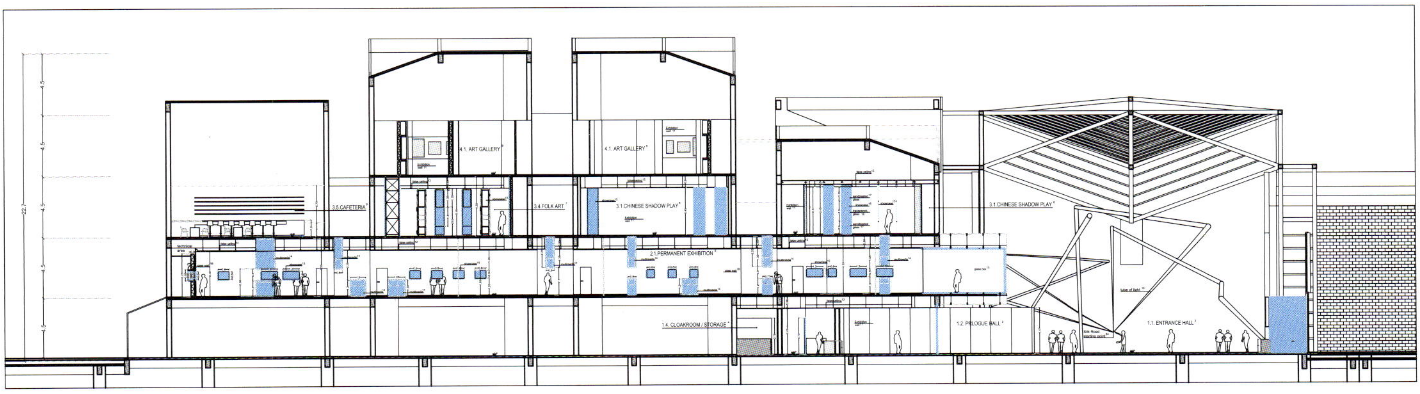
4.1. ART GALLERY
4.1. ART GALLERY
3.5 CAFETERIA
3.4 FOLK ART
3.1 CHINESE SHADOW PLAY
3.1 CHINESE SHADOW PLAY
2.1 PERMANENT EXHIBITION
1.4. CLOAKROOM / STORAGE
1.2. PROLOGUE HALL
1.1. ENTRANCE HALL

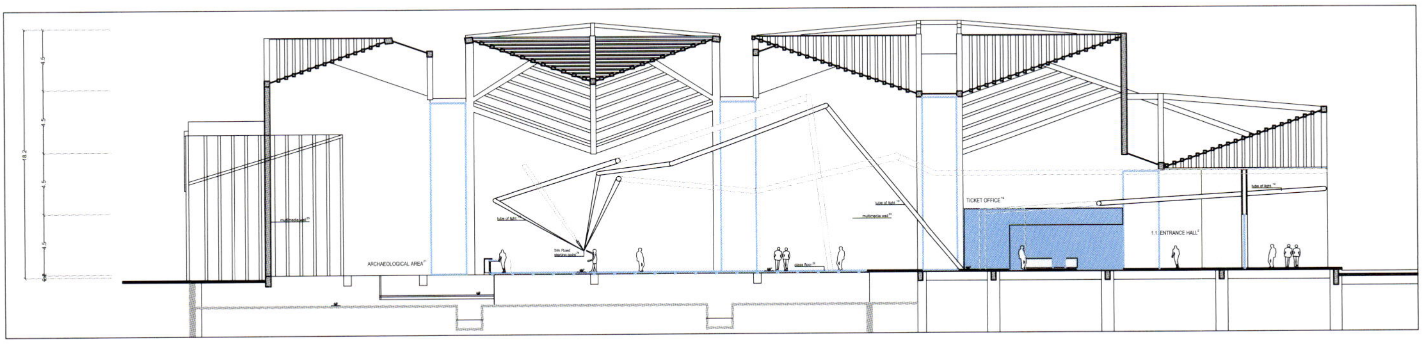
ARCHAEOLOGICAL AREA
TICKET OFFICE
1.1. ENTRANCE HALL

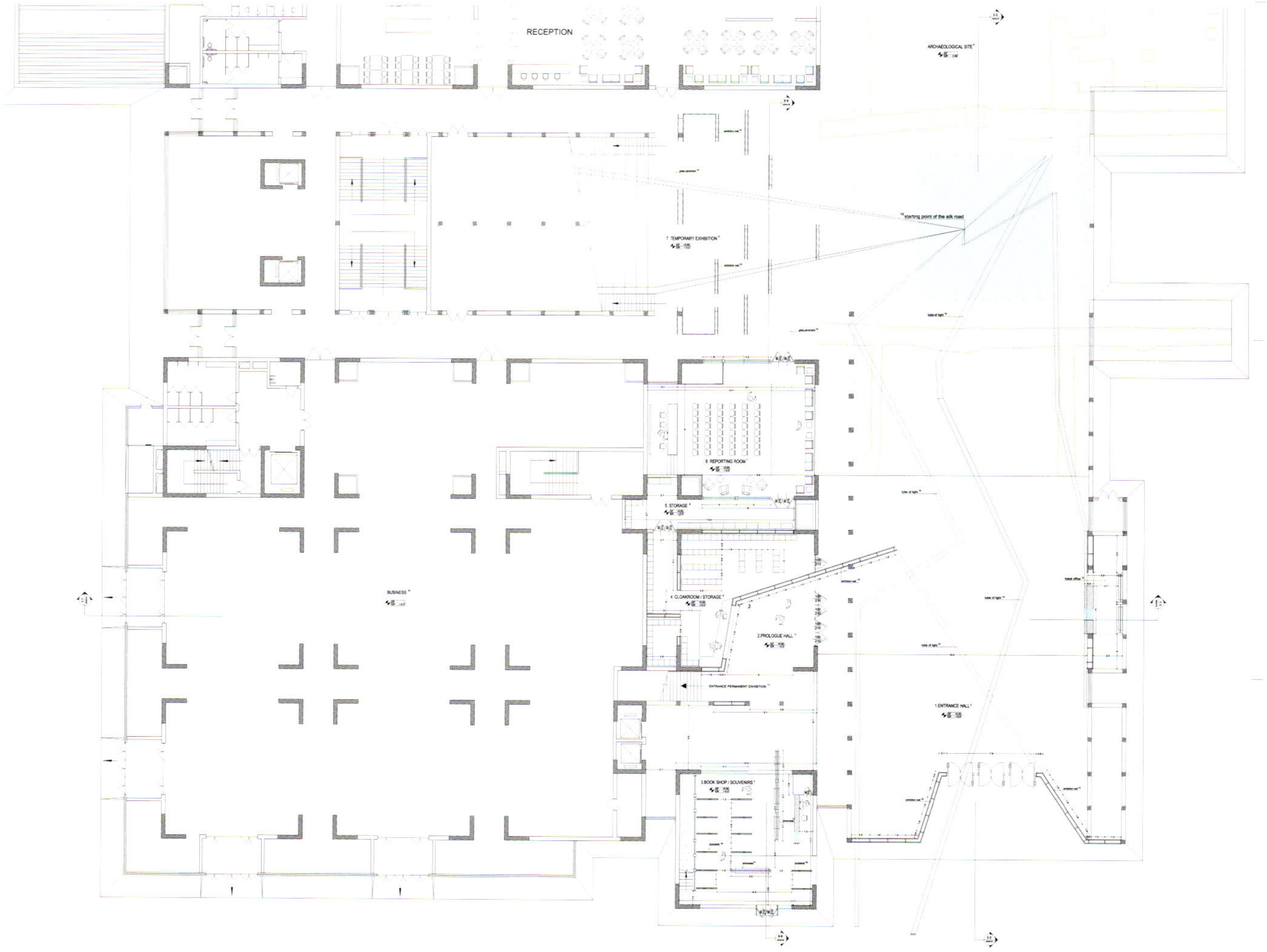

The Tang Dynasty West Market Museum is part of a great development program that aims to re-establish the grandeur of the Tang Dynasty West Market. The museum complex offers a wide range of cultural entertainment including permanent exhibitions, temporary exhibitions, archaeological area and a wide city hall.

The museum complex includes a large public area called City Hall that that corresponds to the ancient road system of Chang'an, capital of the Empire during the Tang Dynasty. A network of light tubes cross the whole space of the City Hall and end in the point considered as the starting point of The Silk Road. These lines of light represent the numerous stages of the Silk Road through centuries, symbolizing its immense variety and complexity. Each line might correspond to the flourishing of the Silk Road during a particular Dynasty, linking the City Hall to the millenarian history of China and its prosperity along the Silk Road.

唐朝西市博物馆是旨在重建唐朝西市宏伟风貌的一项重大开发项目的一部分。此博物馆综合建筑可提供各式各样的文化娱乐设施，包括永久性展区、暂时性展区、考古区和一个宽阔的议事厅。

该博物馆综合建筑内有一个称为议事厅的大型公共区域，与唐朝国都长安的古老城市街道系统相对应。一条由灯管组成的网络贯穿议事厅的整个空间，一直延伸到丝绸之路的起点上。这些灯具组成的线路代表了丝绸之路在很多世纪以来经历的众多阶段，象征着其多样性和复杂性。每条线路均可以对应丝绸之路在某一朝代的辉煌，因而将此议事厅同中国的千年历史及其沿着丝绸之路所创造的辉煌成就联系在了一起。

b.c.1400

b.c.1400

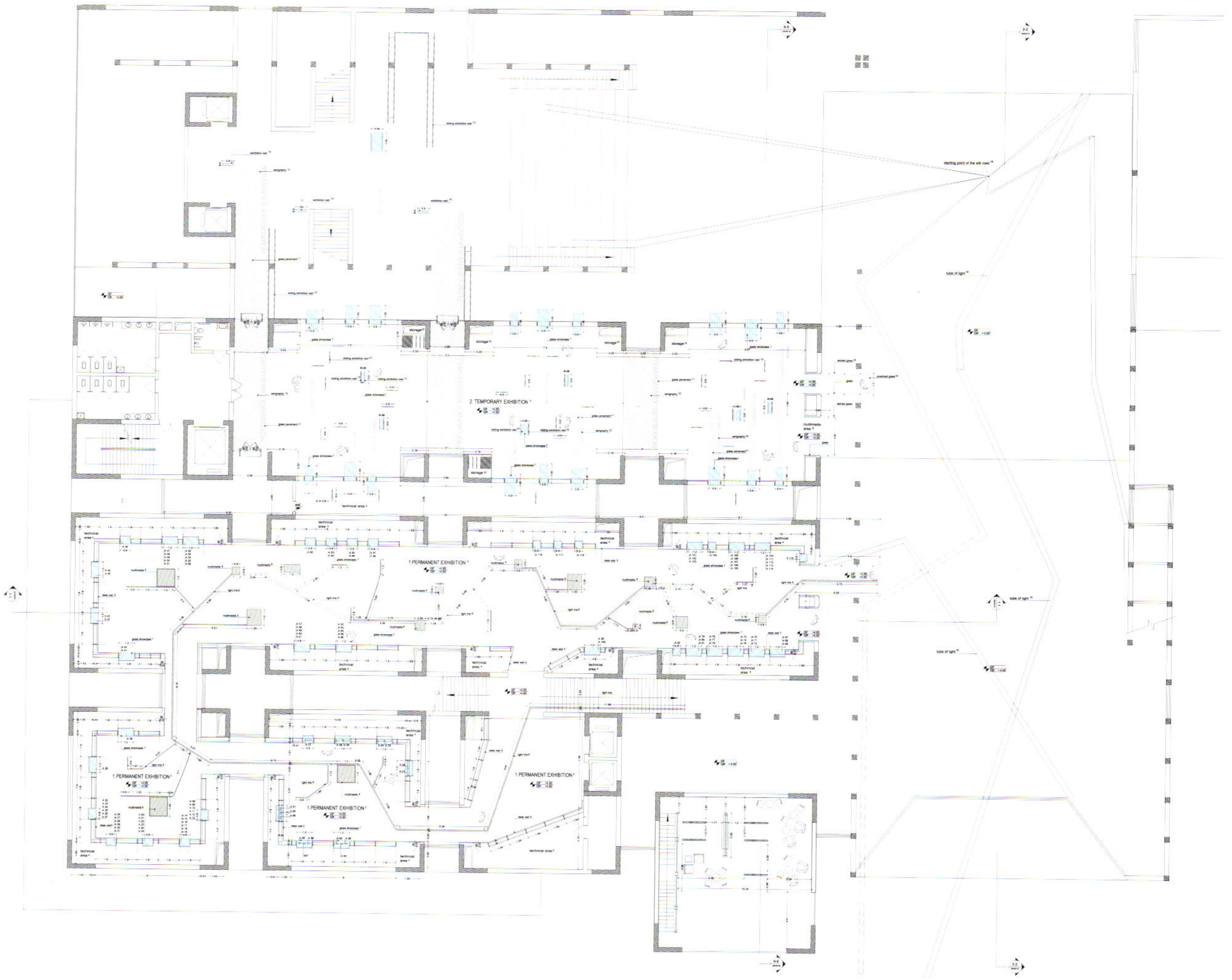

The main permanent exhibition of the Tang Dynasty West Market Museum is dedicated to Tang Dynasty West Market and the Silk Road. This exhibition is the result of a combination of two exhibit levels: artworks on one hand and multimedia on the other. The aim of the design is to offer an incredibly rich cultural proposal, which goes beyond the traditional museum and makes the exhibition unique. The conventional display system is left apart in order to create a hall that provides audience with a complete experience of the exhibition. Along the perimeter walls of the hall the glass showcases are embedded inside a dark steel wall, as virtual windows leaning on the past. Along with the artwork display that runs along the wall, the exhibition present a second level of cultural experience. A line of light representing the abstraction of the Silk Road runs along both ceiling and floor and crosses the exhibition space creating a three-dimensional image. These elements become the physical supports of the Silk Road exhibition, they contain both multimedia and traditional communication that offer a wide experience of the Silk Road.

唐朝西市博物馆的主永久性展区专门用于展示唐朝的西市和丝绸之路。此展区由两层组成：一层展示艺术作品，一层展示多媒体资料。本设计的目的是要提出一种无与伦比的提案，超出传统博物馆的范畴，使展览具有其独特的魅力。它摒弃了传统型的展示系统，旨在营造一座能够使游客完整体验整个展览的大厅的展示系统。沿着大厅周边的墙体，将玻璃展示柜镶嵌入暗色的钢制墙体中，而虚拟窗口将我们同过去紧紧连接到了一起。沿着墙体排列的是艺术品展示区，此项展览为我们提供了另一层文化体验。由灯具组成，沿着天花板和地板排列贯穿整个展示空间的线路代表了丝绸之路的缩影，营造出了一种三维形象。这些成分组成了丝绸之路展览的物理支撑结构，既有多媒体装置又有传统交流工具，使人们能够通过各种各样的模式体验到丝绸之路的魅力。

撒马尔罕
巴比伦
公元前
长安

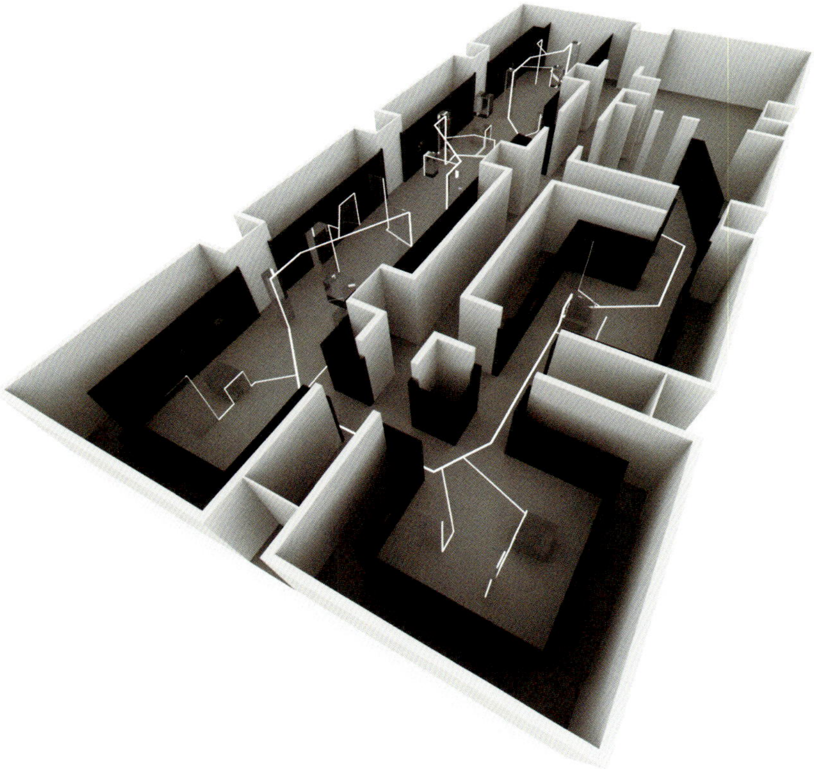

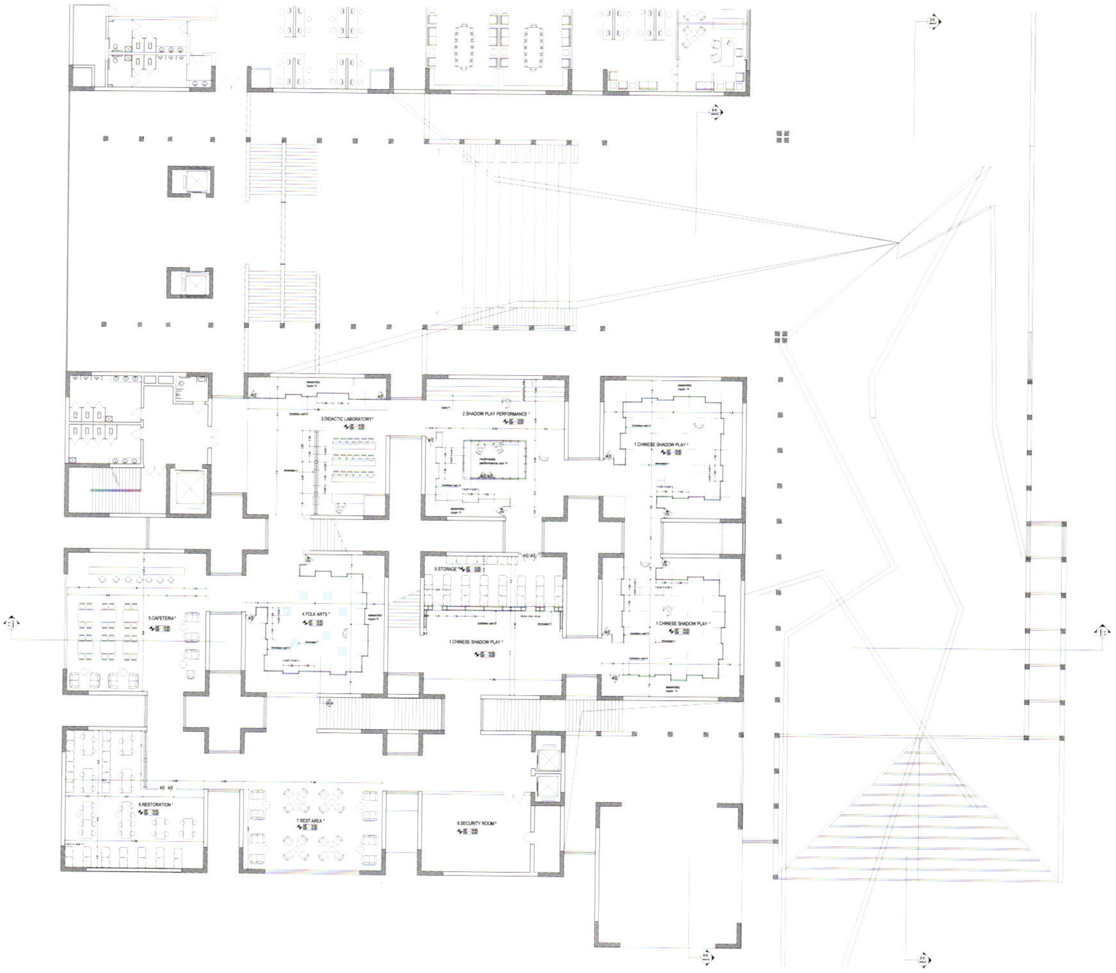

The third floor of the museum is completely dedicated to Shaanxi Shadow Play Exhibition and Folk Art Exhibition. The first section consist of a series of glass vertical panels that go from the floor to the ceiling and contain the original Shadow Plays. These elements can be considered as very slim showcases that keep the artworks safe and visible at the same time. Shadow Plays are set between two extra-clear glass layers anchored both to the floor and the ceiling. The side walls of the exhibition are made of black painted glass panels with white serigraphies representing large details of the shadow plays. The images shown on these panels helps audience to understand the complexity and magnificence of these masterpieces. Besides each glass panel there is a light system that can be activated by visitors simply standing in front of it.

The fourth floor of the Museum complex is totally occupied by an Art Gallery. Keeping the peculiarity of the inclined ceilings of the fourth floor, the exhibition design focuses its attention on the side walls, reshaping the existing ones through the creation of a brand new exhibition wall. This wall is anchored to an apposite steel structure and runs along the perimeter of the hall. In the inner area of the exhibition, at the corner of the halls, there are four rooms that present on one side a large pvc screen printed with big high resolution images representing the geographical, historical or social background of the displayed paintings, on the other side a special multimedia area dedicated to in-dept investigation of the paintings.

博物馆的第三层完全用于展示陕西皮影戏和民间艺术。其第一部分包括一系列的立式玻璃板结构，从地板一直延伸到天花板，其中展出了原汁原味的皮影戏艺术。这部分设置有细长的展示柜，既可以保护艺术品的安全又可以将其展示给世人。皮影戏被设置到连接地板和天花板的两层超洁净玻璃层之间。展厅的侧墙由漆成黑色墙板构成，上面通过百色绢印的模式描述了大量有关皮影戏的详细介绍。这些墙板上所示影像可帮助游客了解这些杰出作品的复杂和伟大之处。在每面玻璃板旁边都有一套灯光系统，游客只要站到前面即可将其打开。

此博物馆综合建筑的第四层完全由美术馆占据。在保持第四层天花板的独特的倾斜设计的同时，此展厅的设计集中在了其侧墙上，通过建造全新的展示墙重塑了原来的墙体。此墙体被锚定在对面的钢制结构上，沿着大厅的周边分布。展厅被分为三部分：顶部和底部部分由漆黑玻璃板制成，中间部分稍微靠后，是容纳所展示的绘画作品的地方。在展区的内部，大厅的角落处有四间屋子，其一侧是一个大型的PVC屏，上面印有高分辨率的图像，介绍了所展示绘画作品的地理、历史或社会背景。另一侧是一种特殊的多媒体展示区，专门用于对这些绘画作品进行深入的研究。

皮影

Museum of Urbanism

城市主义博物馆

International Competition for the New Museum of Architecture and Urban Planning, Salamanca | Design: n!studio Susanna Ferrini, Antonello Stella with Elena Fernandez | Team: Sara Ferazzoli, Angelo Venturi, Stefano Battaglia, Simone Bove, Maria Angela Corsi, Paolo Rossi, Teresa Zanaria | Client: Ministerio de Vivienda | Period: 2007

NUEVA ARQUITECTURA
ESPAÑOLA
LIVING IN MOTION

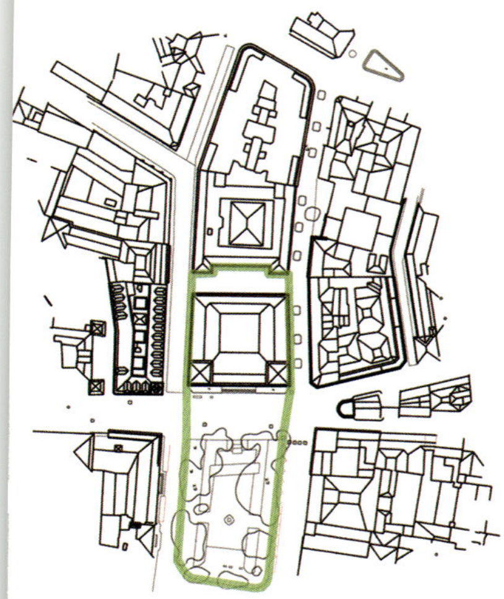

EL **PARQUE** COMO PROLONGACIÓN DEL NUEVO **MUSEO**

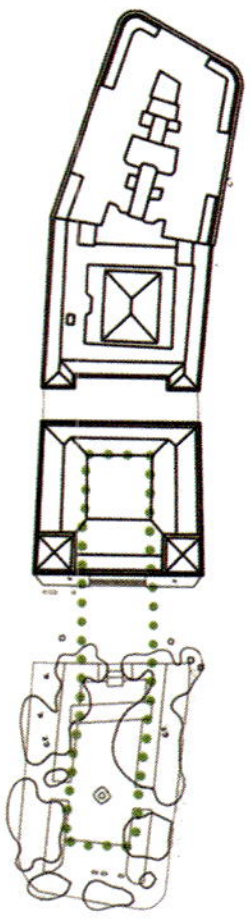

EL **PARQUE** ENCUENTRA

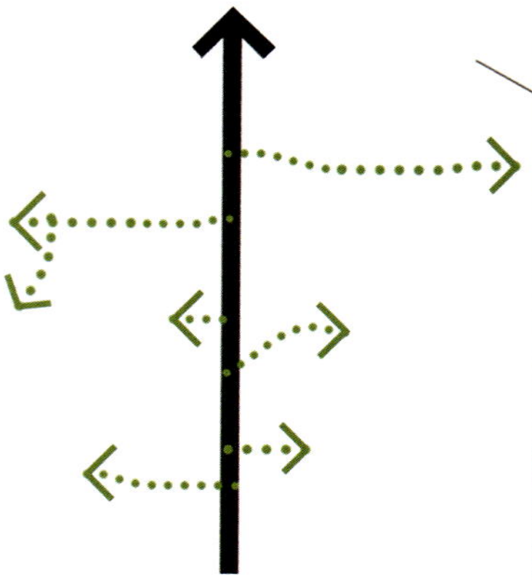

EL **RECORRIDO MUSEÍSTICO** COMO METAFORA DEL PARQUE.

ATRIO PUBLICO

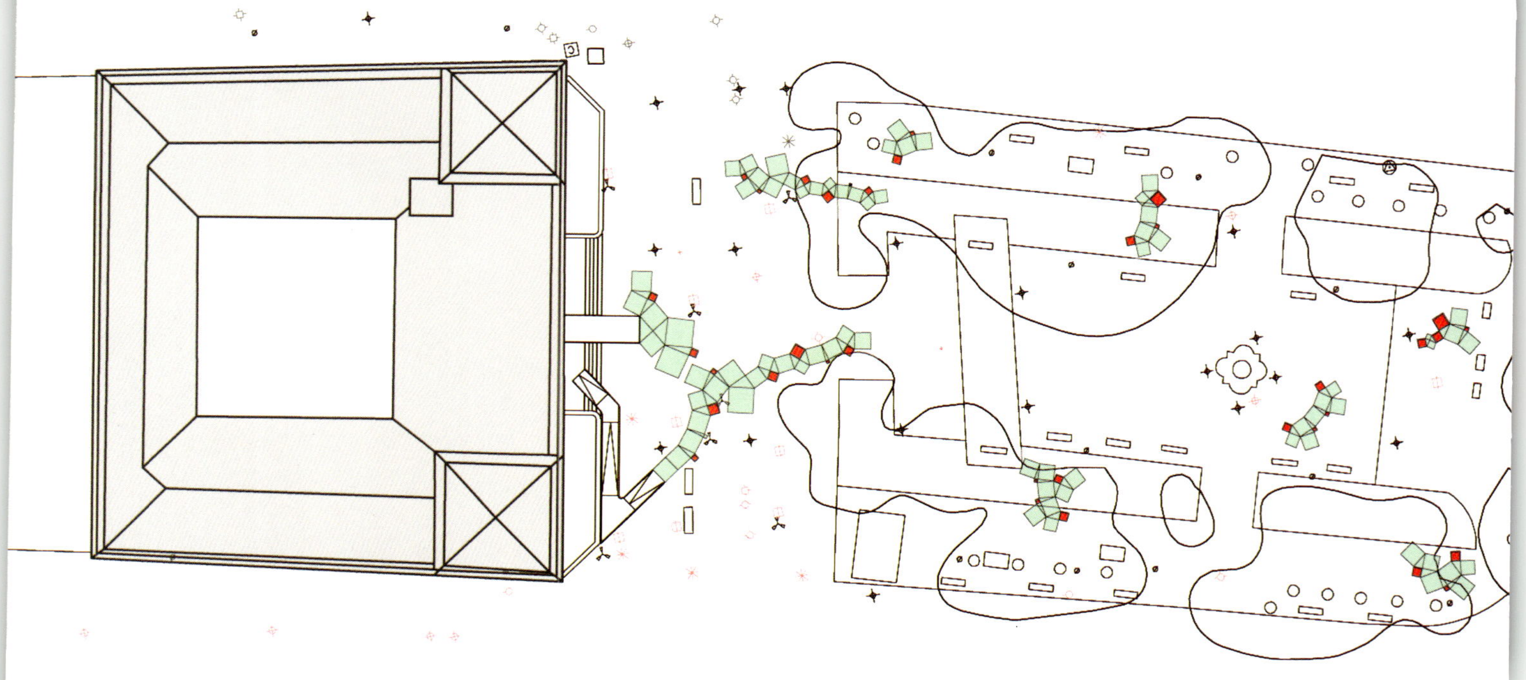

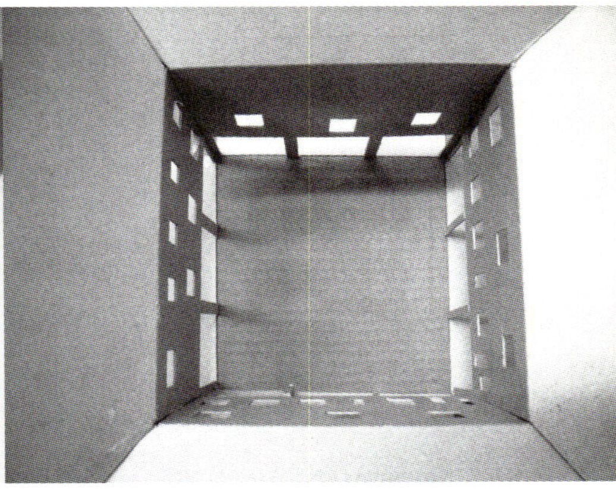

LA IDEA ES LA DE TRATAR EL PATIO COMO
UNA ESTANCIA.

LA CORTE ES UN ESPACIO PUBLICO RELACIONADO
A LA CALLE Y AL PARQUE. ES LA HALL DEL
MUSEO. ES EL ... DE COMUNICACION
VERTICAL, ... LAS INFORMACIONES
... MUSEO Y ES EXPOSICION SI
... DE SU 'PIEL'.
LA 'ESTANCIA DE LA CORTE' ESTÁ PENSADA
HECHA CON UN REVESTIMIENTO ('PIEL') DE LAS
FACHADAS DEL PATIO INTERIOR.
EL SISTEMA PREVEDE BANDAS VERTICALES QUE SE
VAN MOVIENDO PARA ALBERGAR LAS ENTRADAS
DE LA VARIAS SECCIONES EN LAS PLANTAS
PRIMERA Y SECUNDA.
LAS BANDAS GIRAN EN LA CUBIERTA Y EN EL
SUELO.

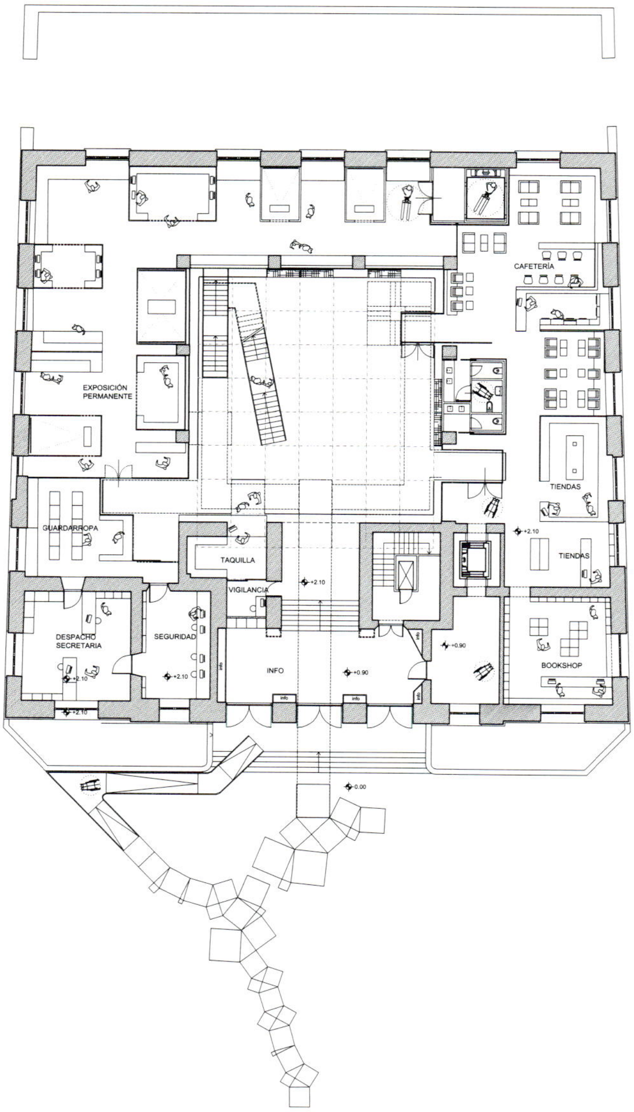
EXPOSICIÓN PERMANENTE
CAFETERÍA
TIENDAS
GUARDARROPA
TAQUILLA
VIGILANCIA
TIENDAS
DESPACHO SECRETARIA
SEGURIDAD
INFO
BOOKSHOP
+2.10
+0.90
0.00

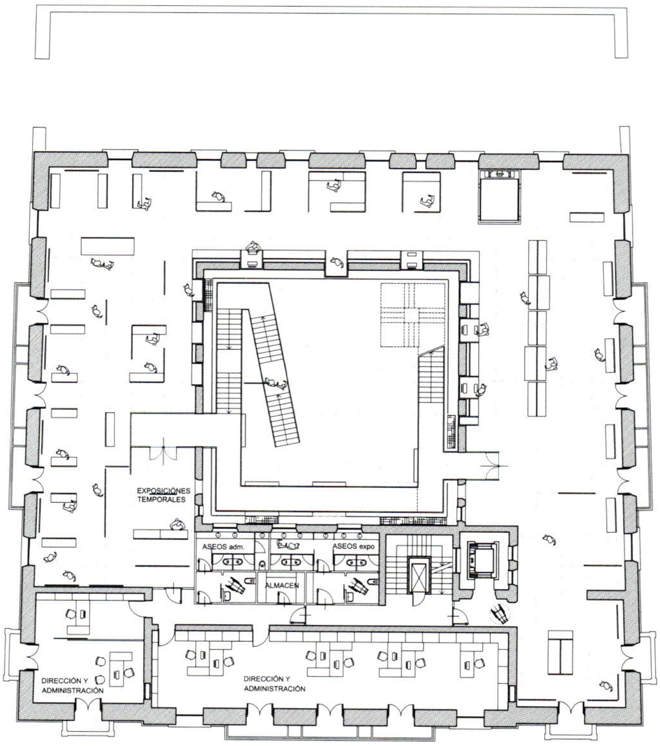
EXPOSICIONES TEMPORALES
ASEOS adm.
ASEOS expo
ALMACEN
DIRECCIÓN Y ADMINISTRACIÓN
DIRECCIÓN Y ADMINISTRACIÓN

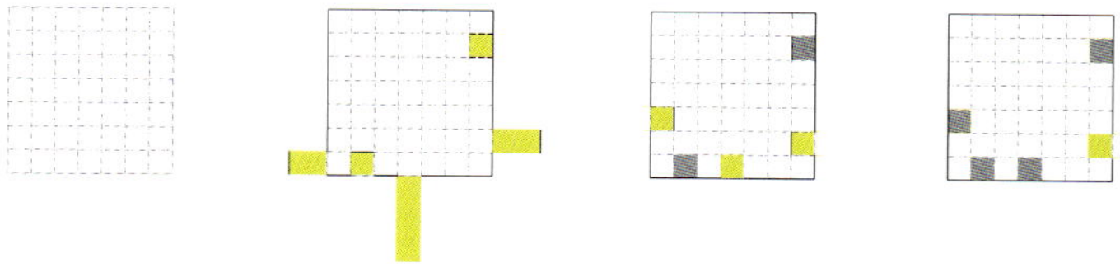

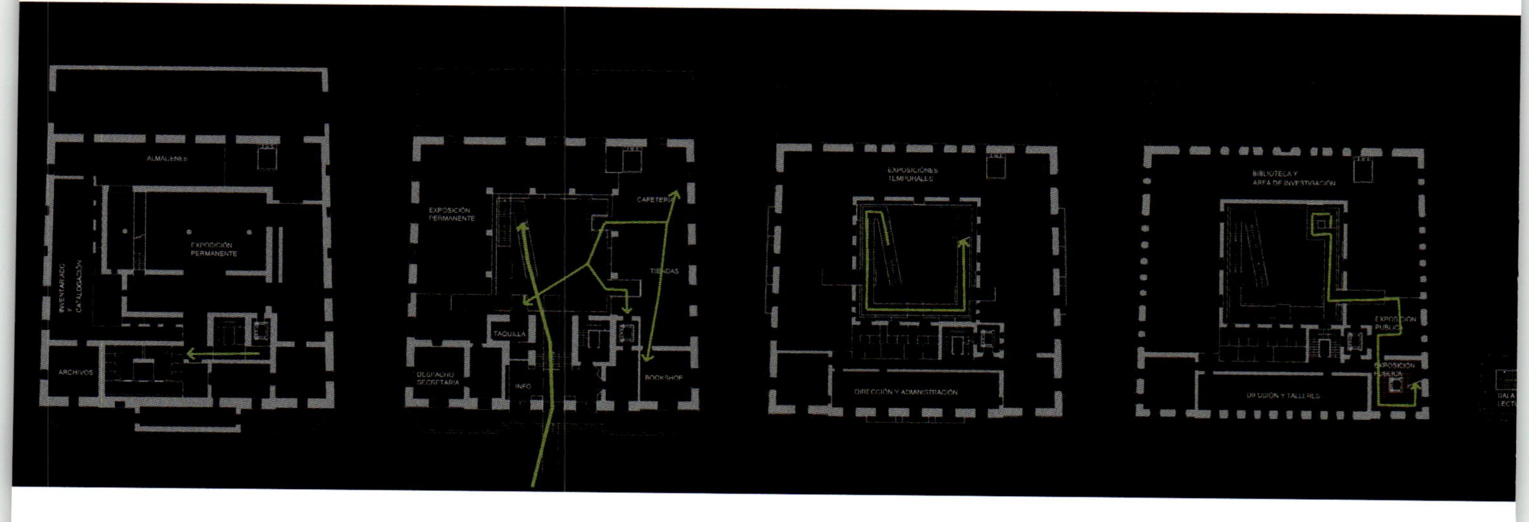
ALMACENES
EXPOSICIÓN PERMANENTE
ARCHIVOS
EXPOSICIÓN PERMANENTE
CAFETERÍA
TIENDAS
TAQUILLA
DESPACHO SECRETARIA
INFO
BOOKSHOP
EXPOSICIONES TEMPORALES
DIRECCIÓN Y ADMINISTRACIÓN

The project proposes a consideration of the internal courtyard as a public space and an extension of the urban space and adjacent park. Similar to a covered plaza, the courtyard is capable of hosting different urban events, such as performances, conferences, concerts, etc., and is equipped with cultural and commercial functions accessible by the city's inhabitants.
Furthermore, the courtyard is also a space of information, used to promote architectural culture at various levels. It thus contains a display path that begins in the plaza at grade and rises up through the building to one of its two towers, offering a suggestive view of the roofs of Salamanca.
The courtyard is like a "room"; it is the hall of the museum, the nucleus of vertical communication, offering information about the museum's activities through its "skin".
The skin that defines the five facades of the volume is composed of vertical strips that move to host the entrances to the various levels of the museum. The elements of vertical communication are thus similar to three-dimensional belts that slip inside the open spaces between the strips of the façade.
The uniform skin, composed of aluminium cells filled with methacrylate elements of varying colour and density, refers back to natural images that can be interpreted at different scales.

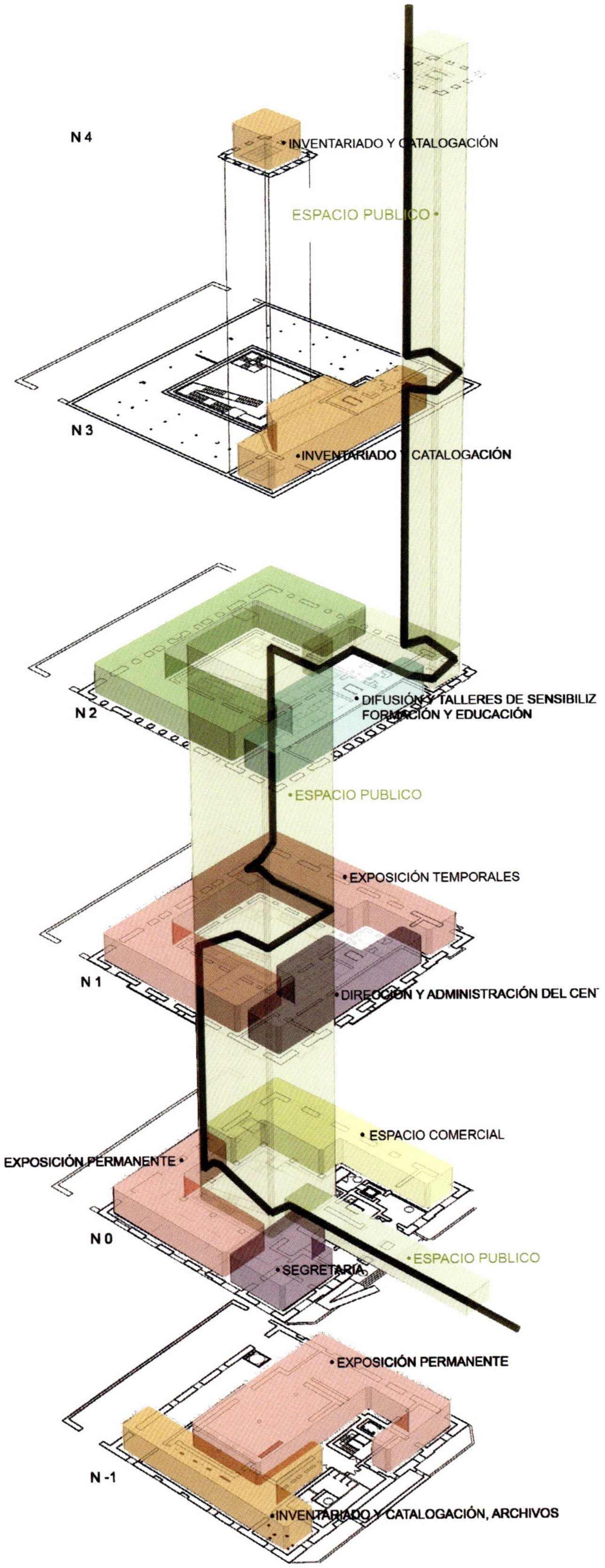

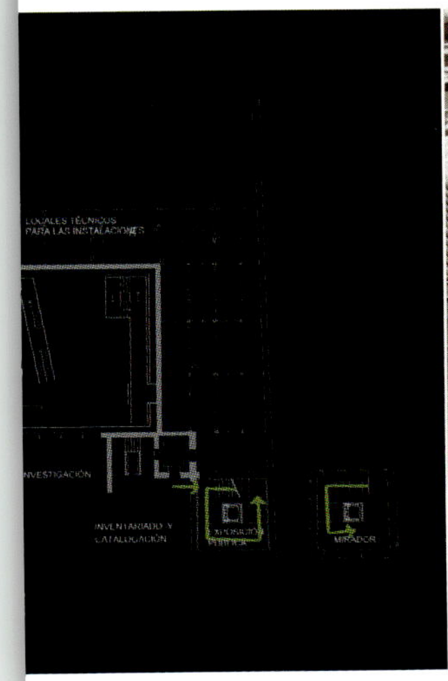

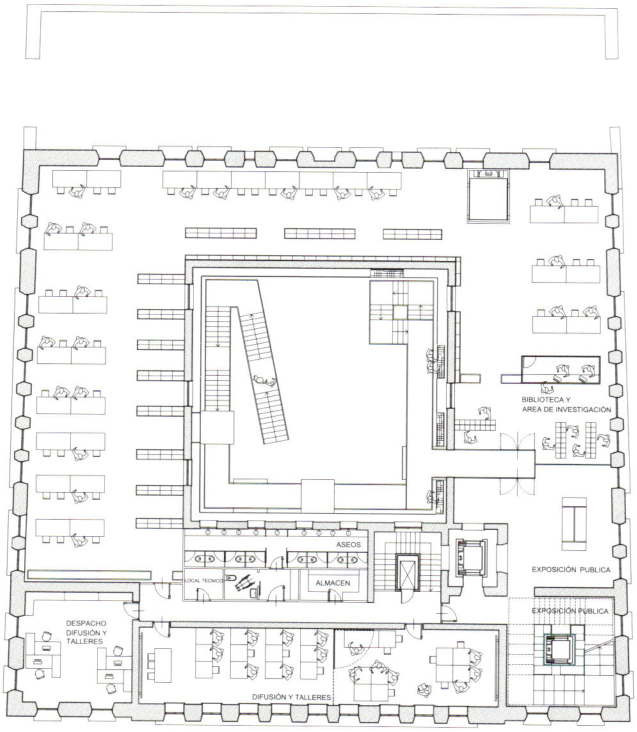

BIBLIOTECA Y
AREA DE INVESTIGACIÓN
ASEOS
LOCAL TECNICO
ALMACEN
EXPOSICIÓN PUBLICA
EXPOSICIÓN PUBLICA
DESPACHO
DIFUSIÓN Y
TALLERES
DIFUSIÓN Y TALLERES

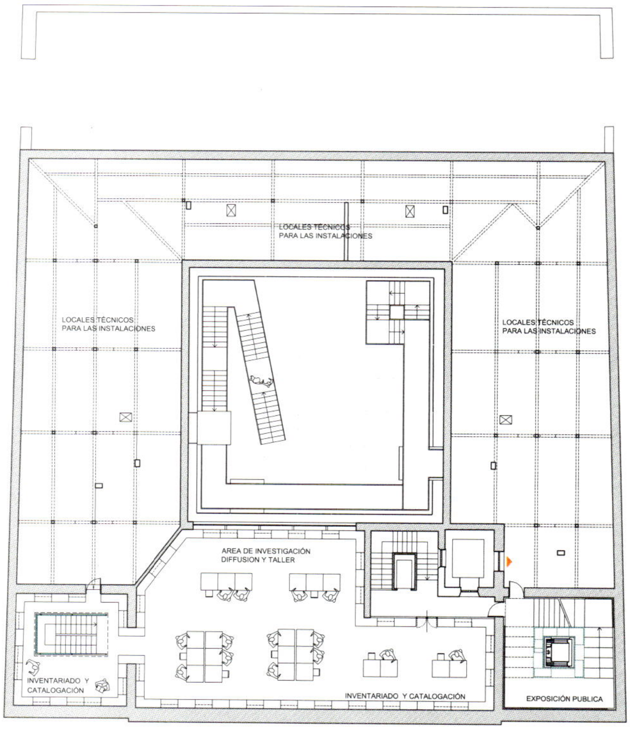

LOCALES TECNICOS
PARA LAS INSTALACIONES
LOCALES TECNICOS
PARA LAS INSTALACIONES
LOCALES TECNICOS
PARA LAS INSTALACIONES
AREA DE INVESTIGACIÓN
DIFFUSION Y TALLER
INVENTARIADO Y
CATALOGACIÓN
INVENTARIADO Y CATALOGACIÓN
EXPOSICIÓN PUBLICA

EXPO
PUBBLICO
BIBLIOTECA
EXPO
TEMPORALES
2
LIVING IN MOTION

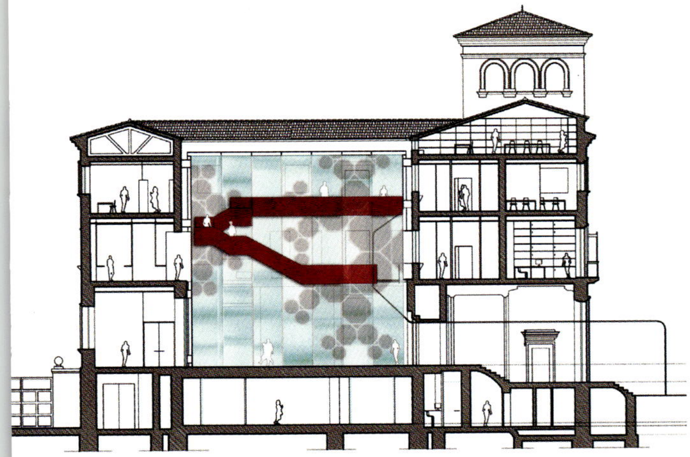

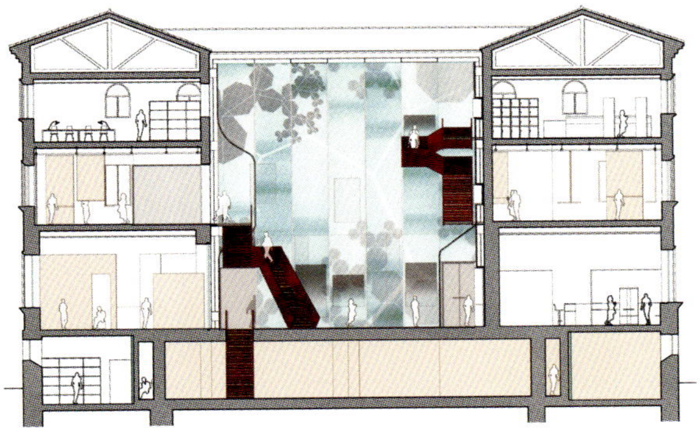

本项目提出了一种将其内部庭院用作公共区域作为城市空间和邻近公园的扩展的设想。与有顶棚的广场类似，在此庭院内可以举办各种不同的城市活动，如表演、会议、音乐会等，并配备有城市居民可以使用的文化和商业功能。

另外，此庭院还是一个获取信息的场地，用于在各个层次促进建筑文化的发展。这样它就具有一个从位于同一水平的广场开始随后上升通过该建筑物通向其两座塔之一的展示通道，使人们可以联想到的Salamanca屋顶。

此庭院就像一间房间一样，是博物馆的大厅，竖直交流的核心区域，通过其“表层”揭示了该博物馆活动的信息。

该建筑体五个外表面结构构成的表层由竖直的带状结构构成，随着其移动即可抵达该博物馆的各层入口。这样各个部分之间的竖直沟通就同一种三维带状结构类似，这种结构插入到了位于外表面带状结构之间的空地之中。

其统一的表层由铝制隔间构成，中间填充有颜色和密度各异的异丁烯酸酯材料，使自然景象在各种尺度上被诠释。

Bari Archaeological Museum

Bari考古博物馆

International Shortlisted Competition for the New Archaeological Museum, Bari - Second Phase | Design: n!studio Susanna Ferrini, Antonello Stella with Kengo Kuma Associated Architects | Team: Simone Bove, Stefano Battaglia, Vito Fortini, Paolo Rossi, Fabio De Carolis, Francesca Fontana, Anna Hollstein, Angelo Venturi, Claudia Campone, Francesco De Tulio | Consultants: Carolina De Camillis, Riccardo Fibbi Lighting Design, Monica Morbidelli Restoration, Isabella Damiani Archaeology, W&media Multimedia | Engineering: 3ti Progetti Italia | Period: 2008 |

heritage

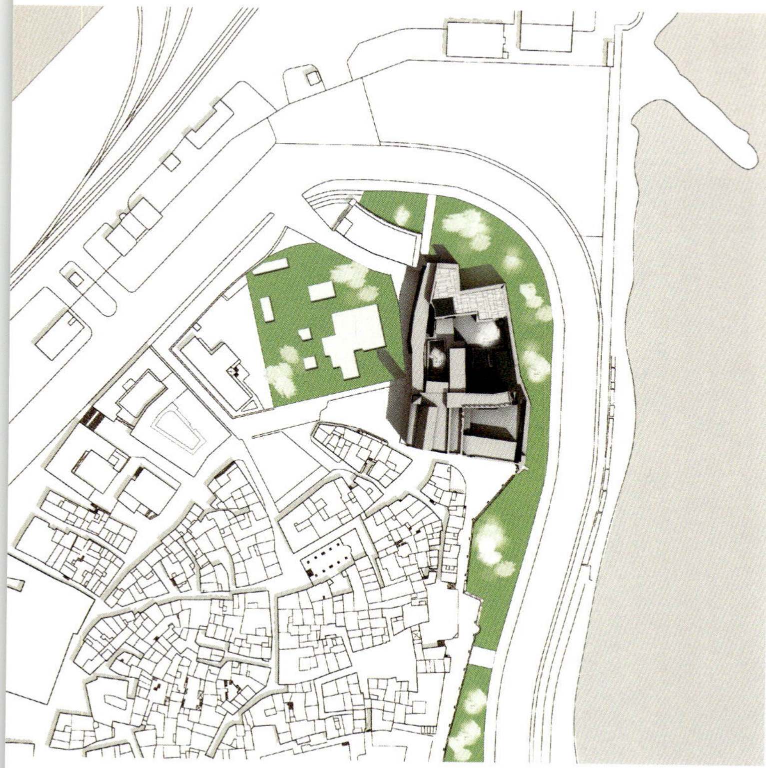

Our proposal is inspired by the significant quality of the existing historical complex, representative of the phases of development of the city of Bari, and the idea of rendering it a central node. The project defines the image of the new cultural centre in the heart of ancient Bari, becoming an element of attraction and point of reference for the entire Mediterranean basin. The new museum complex must become a living and pulsing organism, inserted within the urban context of the historical city and the flows of life in the contemporary city. As part of this approach, the former convent of Santa Scolastica, converted into a museum and centre for temporary exhibitions, no longer acts as a barrier, but as the natural continuation of the path along the walls of the ancient city, extended towards the sea and the East. The project thus interprets the museum complex as an ideal hinge of flows between the ancient city, the waterfront and the port. The project creates an emblematic connection between the museum and the city using the urban path along that runs along the 16th century walls, from the Piazza Venezia and through the museum/former Santa Scolastica convent, connecting with a panoramic terrace. This places significant emphasis and centrality on the space between the two internal courtyards, evidence of the history of the different eras of development of the city. This area is proposed

PORTO
STAZIONE MARITTIMA
P
P
PORTO
STAZIONE AUTOBUS
P
LUNGOMARE
STAZIONE FERROVIARIA

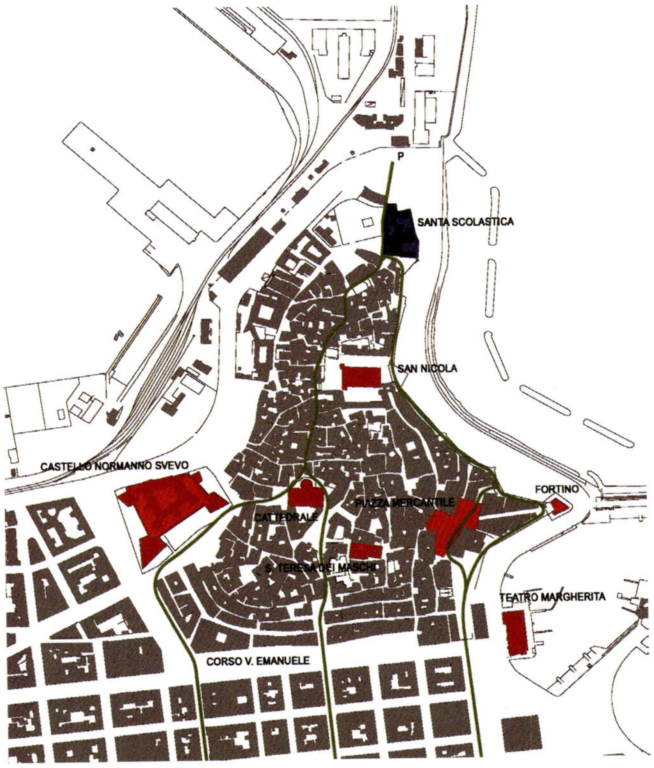
P
SANTA SCOLASTICA
SAN NICOLA
CASTELLO NORMANNO SVEVO
FORTINO
CATTEDRALE
PIAZZA MERCANTILE
S. TERESA DEI MASCHI
TEATRO MARGHERITA
CORSO V. EMANUELE

LA CITTA' IN ETA' ROMANA
PLASTICO INTERATTIVO
CHIESA XI secolo

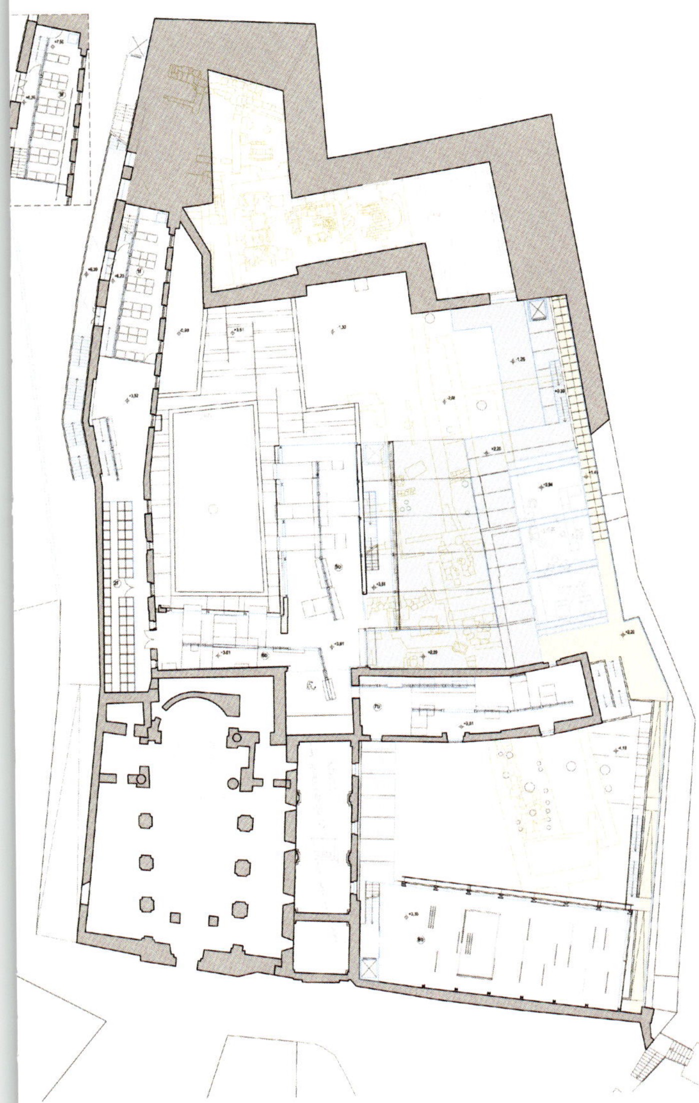

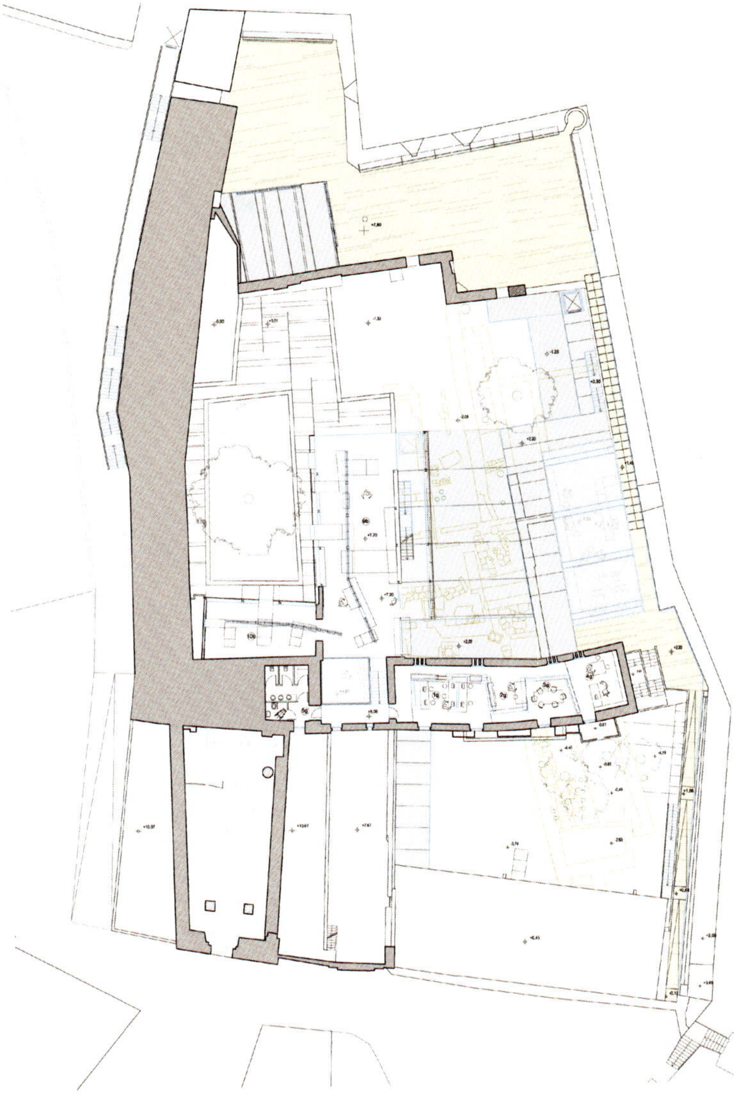

as the ideal space for an open-air museum, focused on the historic evidence of the city of Bari present in the courtyards. The courtyard with the suspended glass floor overlooking the archaeological level is similar to a body of water, as if the sea had reclaimed its original edge, recreating a new neutral level of physical and perceptive continuity. The idea of an archaeological excavation for educational use and exhibitions, a laboratory in the excavation, hypothesises the idea of its musealisation and mise en scene such that it co-exists and is integrated with the idea and fruition of the archaeological museum. Another important factor in this context regards the choice of intervening solely in those structures added in the 1970s: by maintaining the volumes and changing only the skin we re-consider their insertion within the existing historical context. The new facades, clad in slabs of Trani stone are interwoven with the existing as part of a game of chromatic references and geometric matrixes, an intent that is also made clear by their geometric pattern, which alternates with sheets of glass. Our proposed use of materials, some of which are traditional, such as Trani stone, is a contemporary approach to the creation of a relationship with the existing. The less massive treatment of the new facades, differentiated in this way from the existing, can also be read as an expression of the new structure's new function as a museum, giving meaning to the new intervention.

我们所提交的设计的灵感来自该历史性综合建筑现存的部分，其卓越品质代表了Bari市的发展阶段，体现了使其成为一个核心纽带的构思。该项目规划出了位于古老的Bari市中心的新文化中心的形象，使其成为整个地中海盆地地区令人向往和称道之处。该新博物馆综合建筑必须成为一个活生生的、具有脉搏悸动的组织，完全融入了该历史名城的城市规划和现代城市的生活潮流中。作为本项规划的一部分，Santa Scolastica以前的修道院被转化成为了一个博物馆和临时性展览中心，它不再是一道屏障，而是成为了沿着该古老城市城墙自然延伸的一个通道，一直通到大海及东部地区。因而此项目使此博物馆综合建筑变成了连接该古城、海滨地区以及港口地区的理想纽带。此项目通过沿着其16世纪城墙走向的城市通道在博物馆和城市之间创造出了一种标志性连接，从Piazza Venezia开始穿过博物馆（前Santa Scolastica修道院），形成了一条贯通的全景长廊。它着重突出了其两座内部庭院之间的区域的重要性及其核心地位，体现了城市发展不同时代的历史特色。根据我们的提案，此区域是建造开放型博物馆的理想场地，可集中体现这些庭院中Bari城市的历史痕迹。我们设计了耸出考古层之上有着悬浮玻璃地板的庭院，与水体类似，就像是大海重新返回了其最初的边缘，形成了保证物理和感知两方面的连续性的中间层。这种将考古发掘文物现场用于教育和展览的构思，与建设考古博物馆的构思和成果共存并融为一体。我们决定只在20世纪70年代增加的建筑中采取措施：通过维持其原空间体积而只改变其表层，我们重新考虑了将其插入现有历史背景中的模式。其新的正面结构，附上Trani石板层同现有结构交织在一起，就像一种色丰富的几何结构排列的游戏，其几何形状同玻璃板相互交错，也清晰地表达出了其预期用途。我们所提议使用的材料，有些是传统型材料，如Trani石，其整体设计利用是一种可与现有建筑建立关系的当代方法。其新的正面结构经过处理后不再那么雄伟，通过这种方式也将其与现有的建筑区分开来，这也可以被当作是此新建筑作为博物馆的新功能的表达方式，赋予了新的意义。

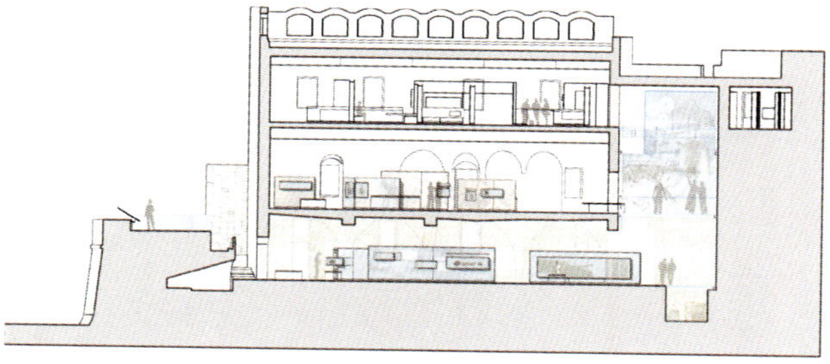

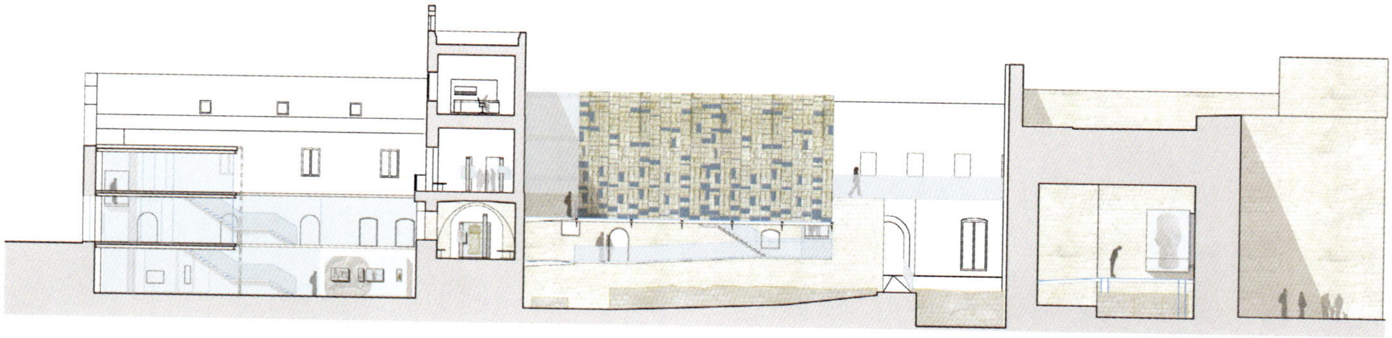

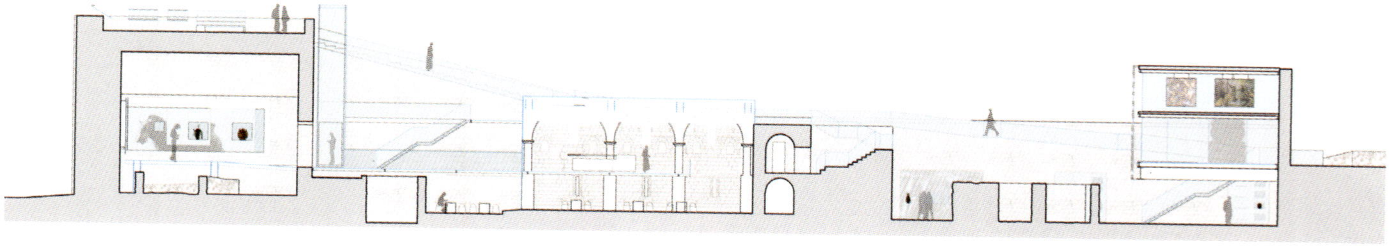

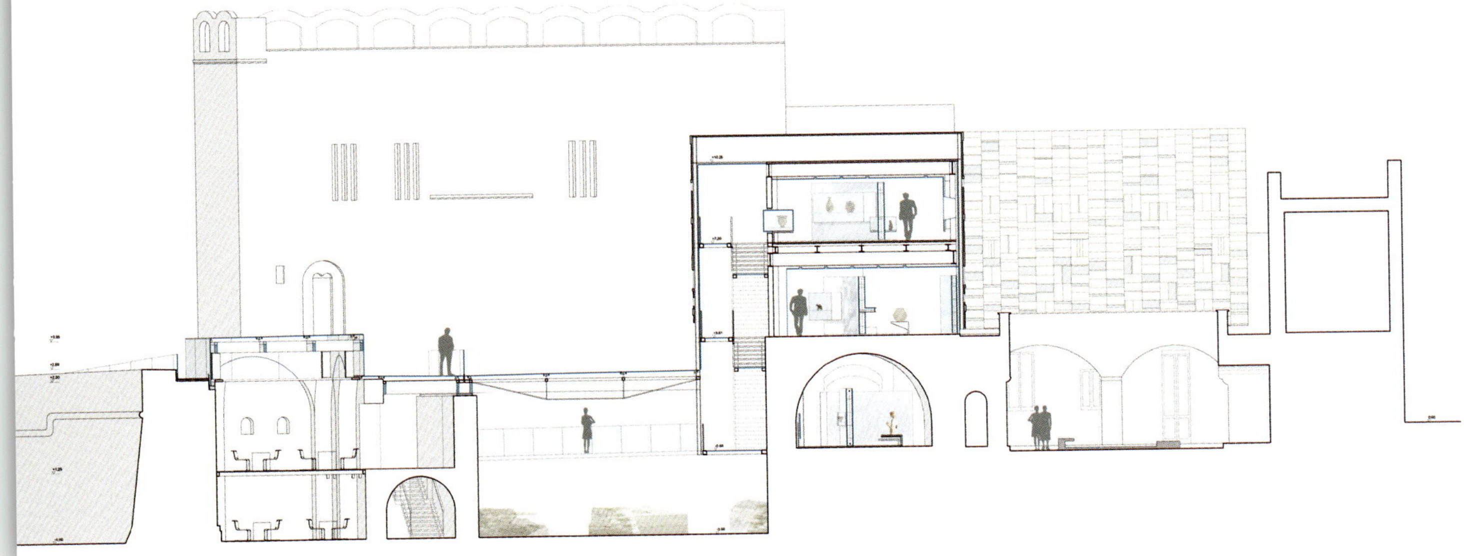

Palazzo Bevilacqua-Costabili

Costabili-bevilacqua改建项目

University of Ferrara School of Economics New Head Office inside Palazzo Costabili-Bevilacqua (16th Century), Ferrara
Design: Antonello Stella, with Stefano Cornacchini, Paolo Iotti | Team: Roberto Meschini, Andrea Pasquato | Client: University of Ferrara | Period: 2006

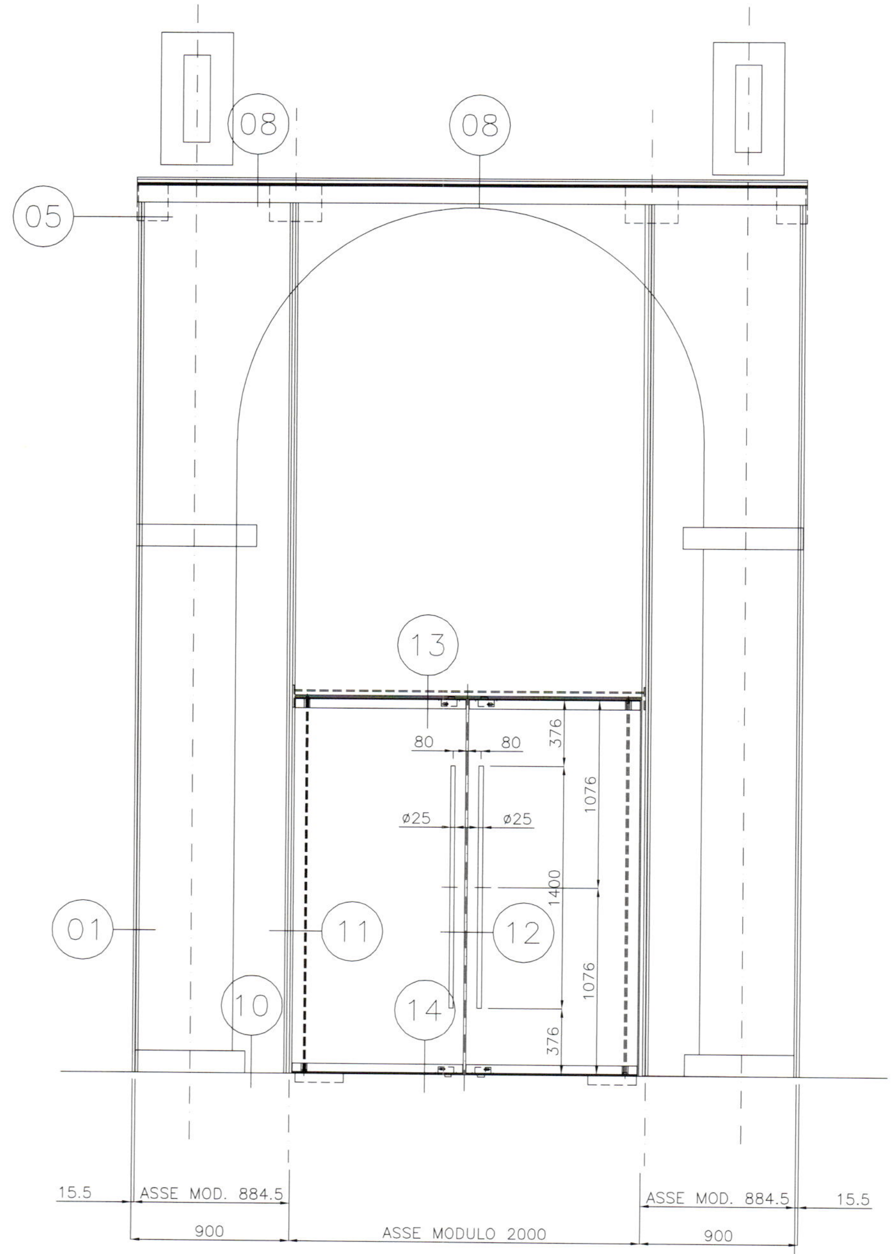
08
08
05
13
80
80
376
1076
ø25
ø25
1400
01
11
12
10
14
1076
376
15.5
ASSE MOD. 884.5
ASSE MOD. 884.5
15.5
900
ASSE MODULO 2000
900

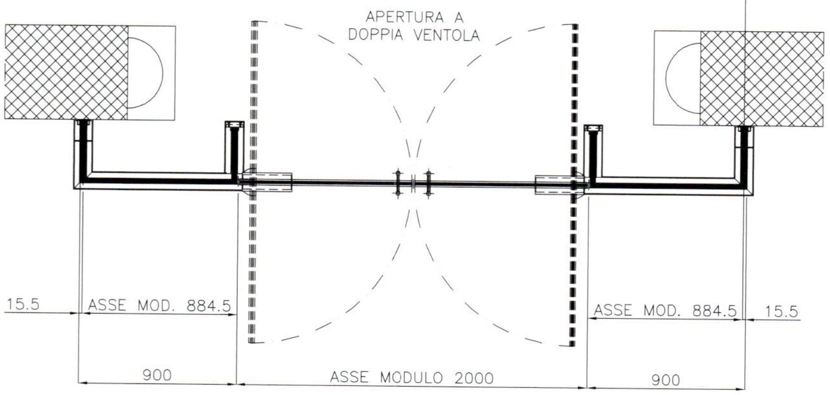
APERTURA A
DOPPIA VENTOLA
15.5
ASSE MOD. 884.5
ASSE MOD. 884.5
15.5
900
ASSE MODULO 2000
900

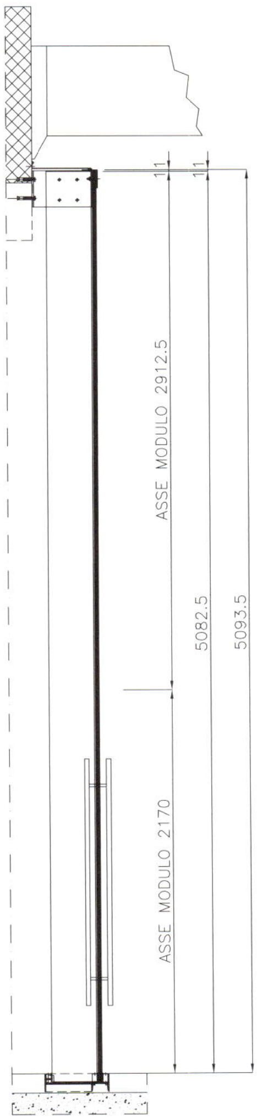
11
11
ASSE MODULO 2912.5
5082.5
5093.5
ASSE MODULO 2170

The design of the new home of the University of Ferrara's Faculty of Economics began with the renovation of the Palazzo Costabili-Bevilacqua, a historic 17th century building with various additions located in the heart of the city of Ferrara. The intervention called for the creation of classrooms, a library and administrative offices and the reuse of the building's two courtyards. The primary new building material is glass, used here as a structural element. This can be seen in the large window on the ground floor that separates the portico from the courtyard: the choice not to use any steel fittings ensures that it does not 'compete' with the historical structures, rendering the contemporary intervention as minimal as possible. Even the technical volume for the electrical systems located in the interior courtyard is clad entirely in backlit, opalescent glass. Historical renovation and contemporary transformation come together as part of an important dialogue based on the clear distinction between historical pre-existences and new interventions. This approach minimises the intention of form and confrontation with the architectural qualities of the existing building. The material and highly decorated consistency of the historical structure is juxtaposed with the lightness and transparency/translucence of the new glass structures, stripped of any decorative elements.

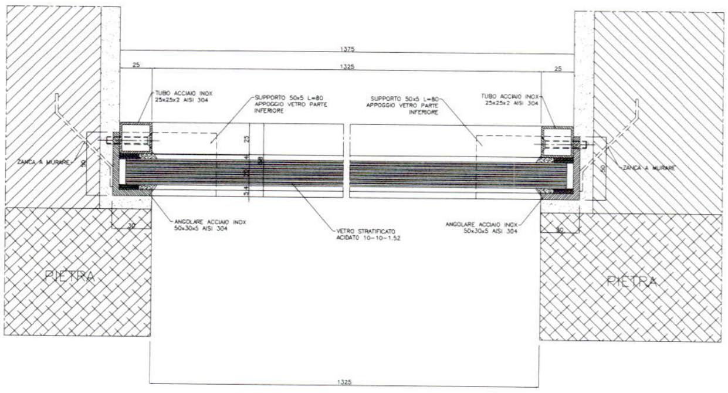

Ferrara大学经济学院教职员工的新家设计就从Palazzo Costabili-Bevilacqua的翻新工程开始，这是一座位于Ferrara中心的17世纪的历史性的、有着各种附属功能的建筑。此项设计需要建造一些教室、一座图书馆、管理办公室和重新利用该座建筑的两个庭院。其所采用的主要新建筑材料是玻璃，在此用作建筑构造中的成分。这可以从底层将门廊同庭院分开的大型窗户所用的材料看出：选择不采用任何钢制配件保证了它不会同那些历史性建筑的结构形成“竞争”，尽可能地减少当代元素的介入成分。即使是位于内部庭院的电气系统技术构件也是完全由背投式乳白色玻璃覆盖在表层而构成。其历史性建筑和新介入建筑之间有着明显的差别，历史翻修和当代转变交融在一起，实现了双方的对话。这种方法最大程度地减少了形式的目的以及与已有建筑之间所发生的冲突。历史性建筑结构的连贯性与新玻璃结构的轻盈和透明或半透明特性并驾齐驱，协调统一。

Ss. Quattro Coronati Library

Ss Quattro Coronati图书馆

New Library inside Ss. Quattro Coronati Monastery, Rome | Design: n!studio Susanna Ferrini, Antonello Stella I Team: Simone Bove, Lara R. Perez-porro | Restoration Team: Prof. Giovanni Carbonara Supervior, Prof. Lia Barelli, Arch. Mara Falconi, Arch. Elisagetta Giorgi, Arch. Monica Morbidelli, Dott. Raffaele Pugliese Archaeologist | Engineering: Prof. Francesco Piccarreta | Carolina De Camillis Riccardo Fibbi Lighting Design Costruction Company: Officine Ragnini S.r.l. Steel and Glass Library, Solema Costruzioni General Restoration, Edi.lux Sas Electrical System Francesca Matera Restorer | Client: Ss. Quattro Coronati Monastery, Community of Augustinian Nuns, Mons. Giovanni Tani Rector of the Monastery

Sit vobis anima una et cor unum in

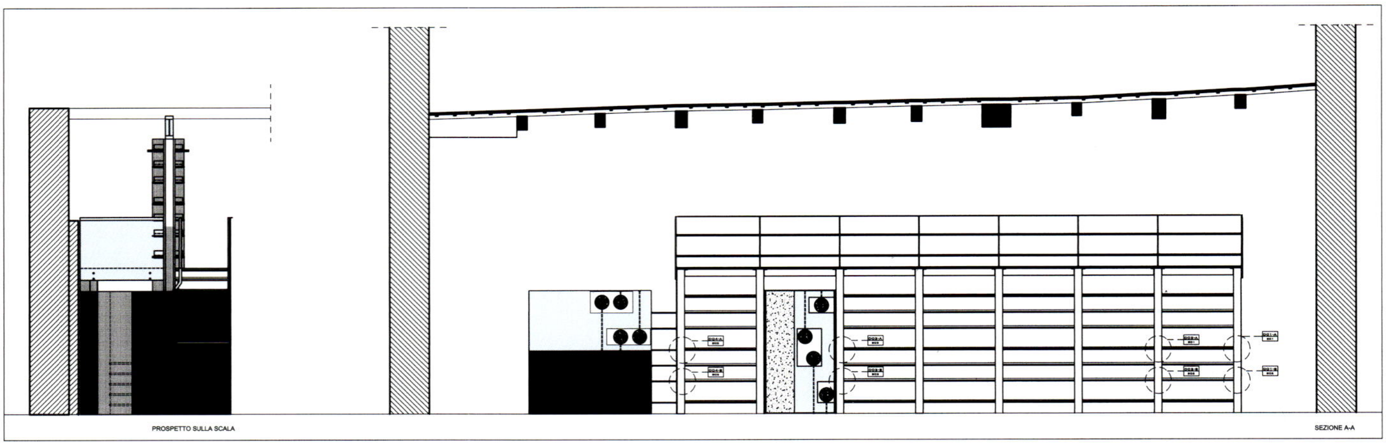

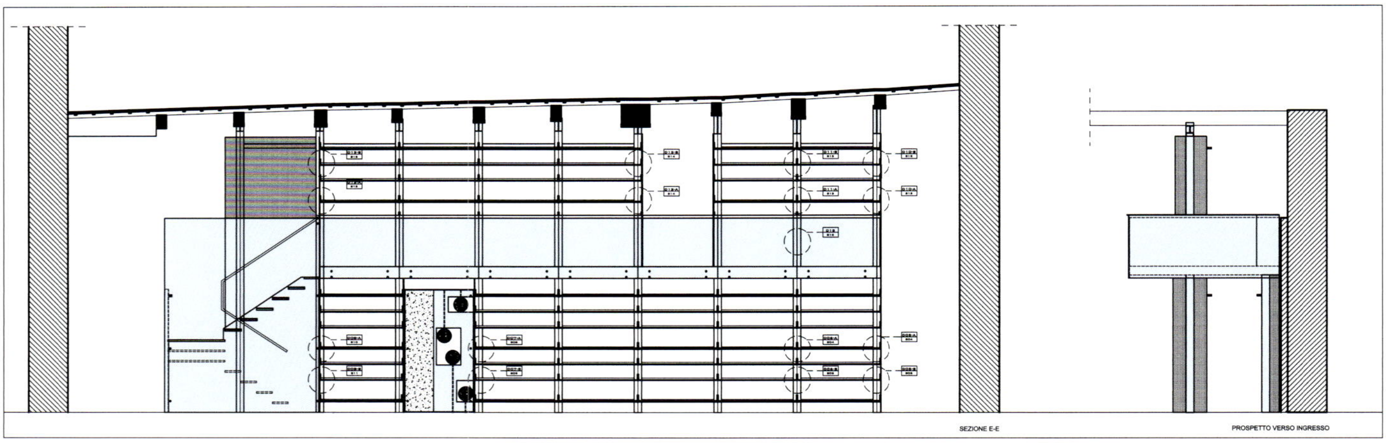

The intervention consists of a new library for the Augustinian Nuns Community inside Ss. Quattro Coronati Monastery in Rome. The project is part of a wider restoration programme of the monastery lead by Prof. Arch. Giovanni Carbonara. The new library accomplishes an innovative double task of displaying books and consolidating the existing wooden beams roof at the same time. As a matter of fact the structure consists of a series of vertical steel HeB profiles corresponding to the roof beams and horizontal "T" and "L" shape steel profiles to support books. Shelves are made of tempered glass and perforated steel mesh. The library is developed on two levels, separated by a 28 mm. temperate opal glass flooring solution and connected by staircase made of 28 mm. temperate opal glass steps free for any steel frame. The staircase elevation consists of opal glass panels in continuity with the first floor glass rail, with serigraphies recalling quotations from the Scriptures. Besides the consulting function the library includes some small exhibition spaces to display archaeological findings coming from the contiguous cloister.

此设计由位于罗马的Ss. Quattro Coronati修道院内的Augustinian Nuns Community新图书馆组成。此项目属于由Prof. Arch. Giovanni Carbonara领导的对此修道院进行修复的项目的一部分。

此新图书馆的设计担负着双重任务，展示书籍和加固现有的木质横梁屋顶。事实上，该建筑结构包括一系列与屋顶横梁相对应的竖直钢质HeB轮廓构造，以及用于支撑书本的水平“T”形和“L”形钢质轮廓构造。书架由钢化玻璃和带孔眼的钢筛构成。

该博物馆为两层，由28mm的耐高温乳色玻璃地板将两层分开，两层之间由耐高温乳色玻璃楼梯连接，没有使用任何钢质支架。楼梯由直接从第一层的玻璃围栏连续延伸出的乳色玻璃板构成，上面还有记录了圣经诤言的绢网印版画。除了可进行咨询活动之外，此博物馆中还设有一些小的展区，用于展示从邻近修道院收集来的考古文物。

verim me, noverim Te
Io sono la via, la verità e la vita

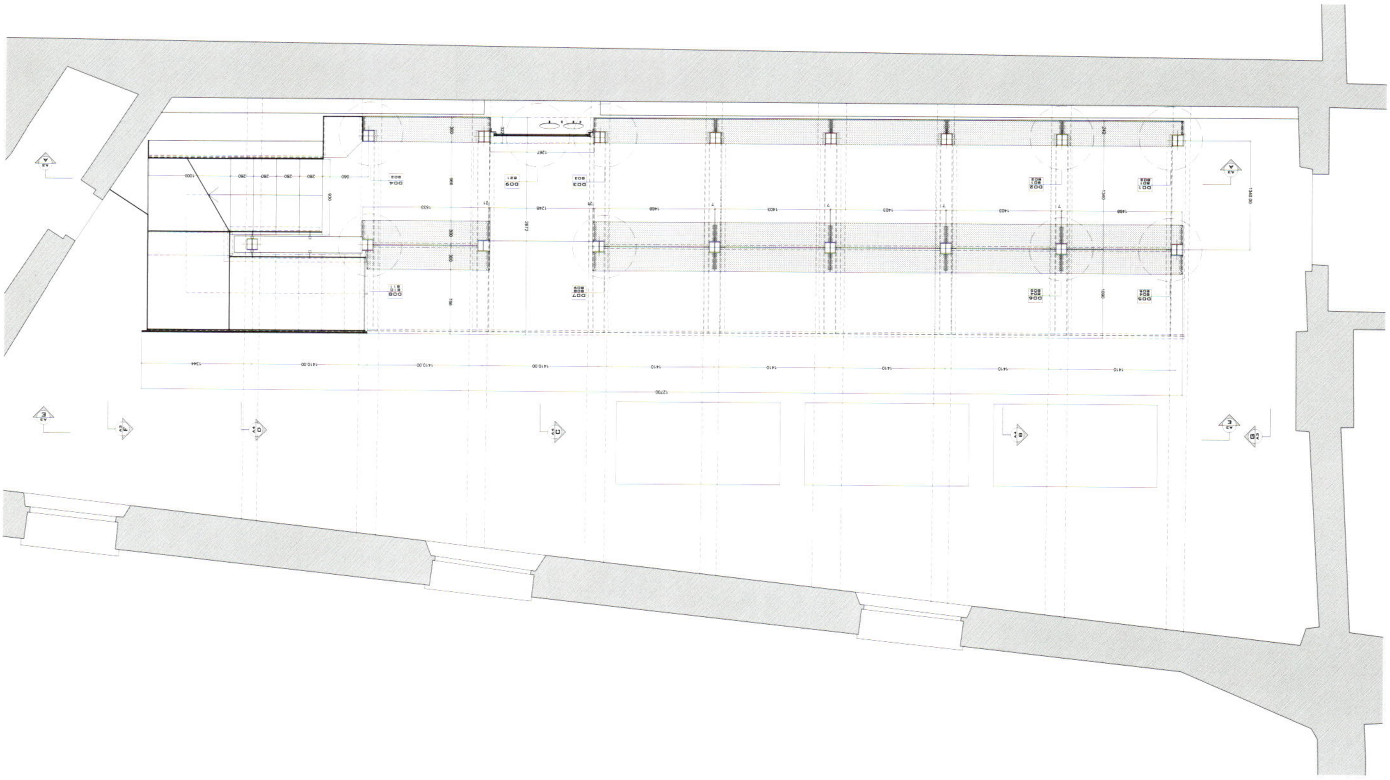

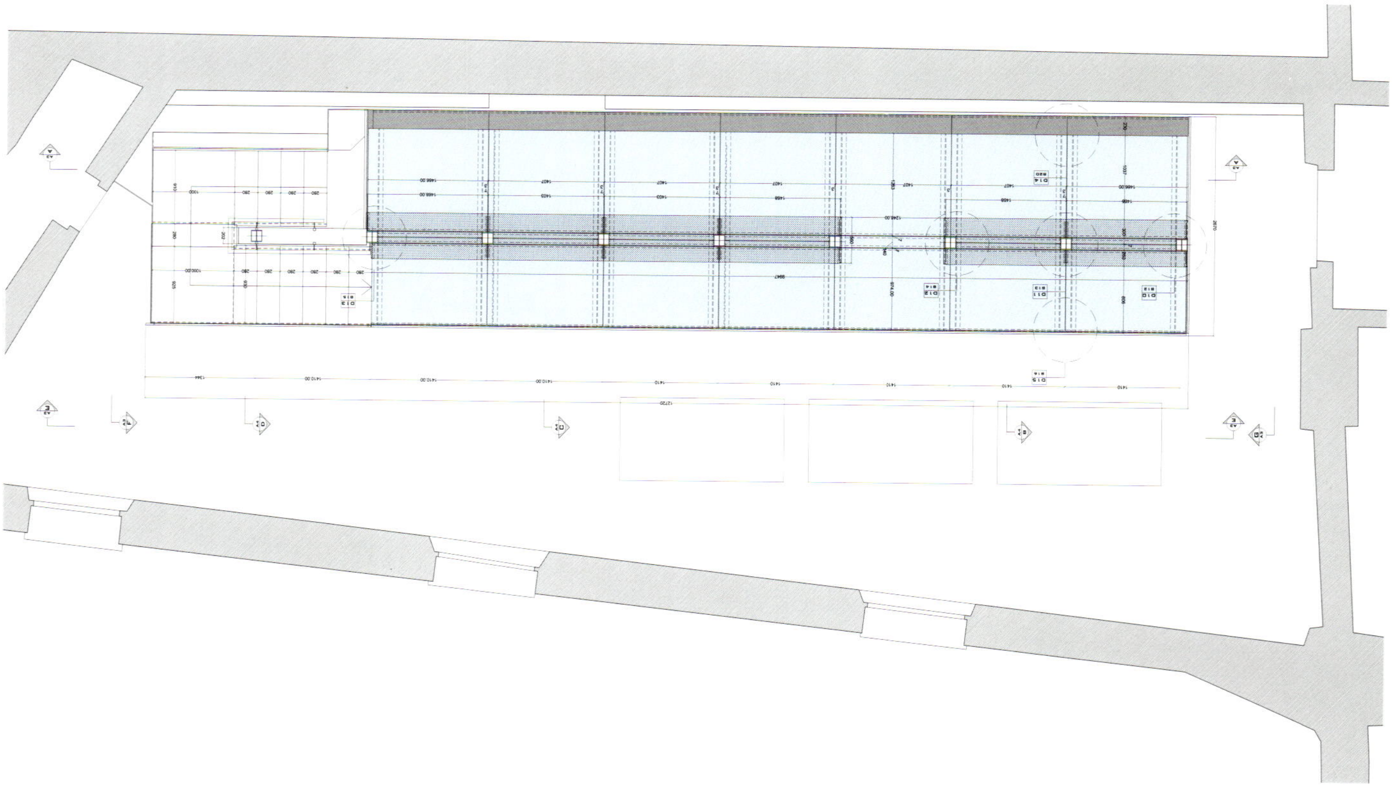

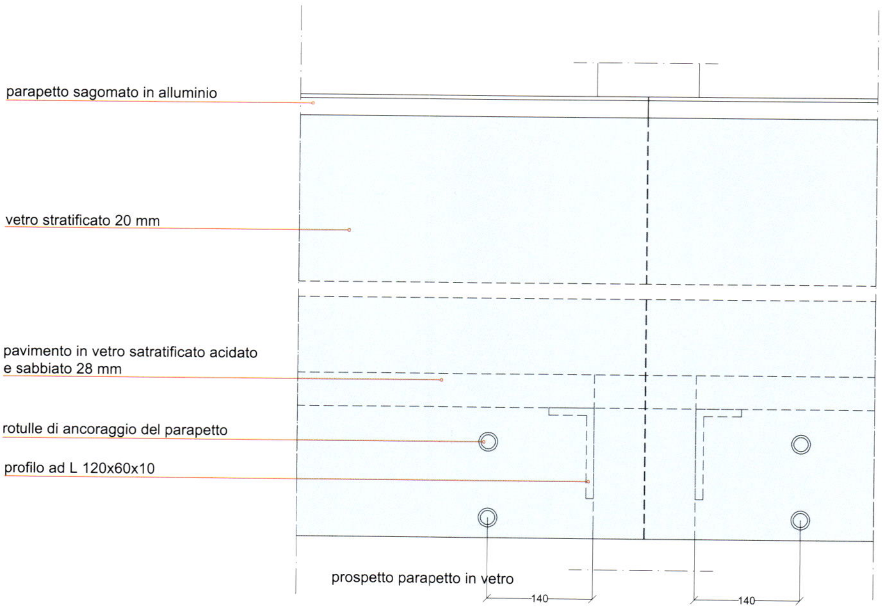

prospetto parapetto in vetro

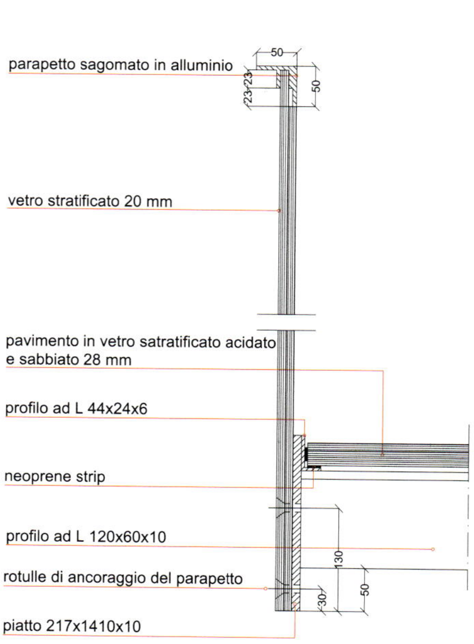

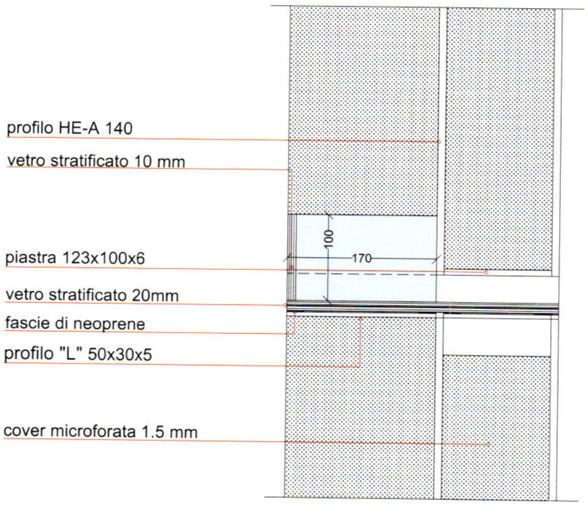

sezione c-c' mensola in vetro

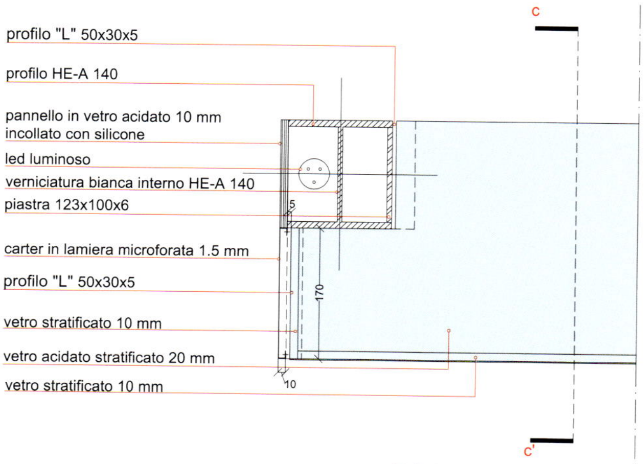

mensola tipo D **D01_B**

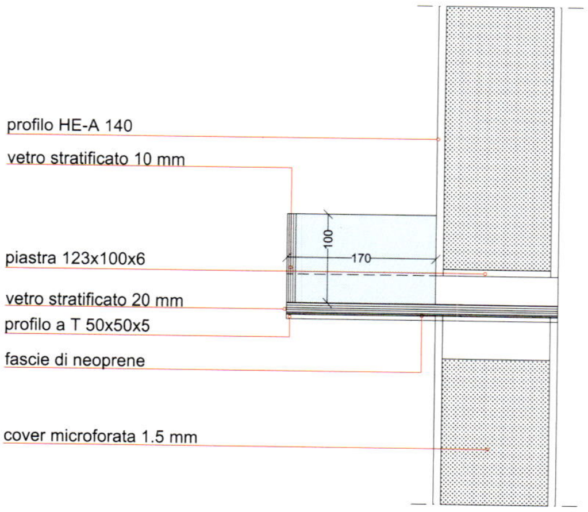

sezione c-c' mensola in vetro

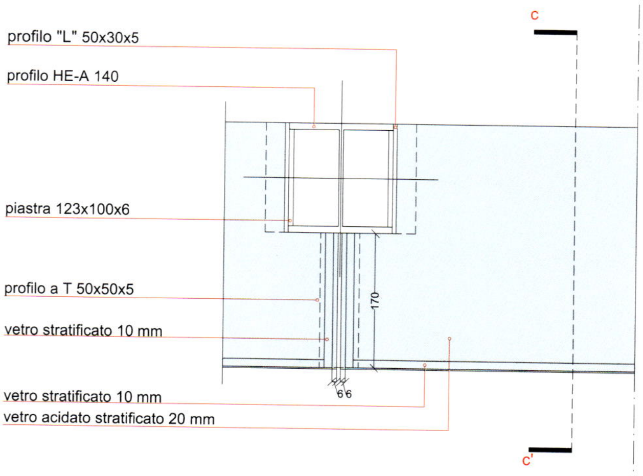

mensola tipo E **D02_B**

In Cristo sono nascosti tutti i tesori della

heritage

Pitigliano Exhibition Centre

tigliano展览中心

novation of the Orsini Fortress, Pitigliano | Design: n!studio Susanna Ferrini, Antonello Stella | Team: Simone Bove, Stefano Battaglia, Gabriele so, Giulia Scaglietta | Consultants: Carolina De Camillis, Riccardo Fibbi Lighting Design | Client: Pitigliano Municipality (Fc) Contructor: E.c.m. . Viterbo General Restoration, Officine Ragnini S.r.l. Pitigliano Exhibition Setting | Period: 2007

Within the historical texture of Pitigliano, the design of an exhibition street proposes an urban path set inside the ancient complex of the Rocca Orsini. The design aims to underline the continuity of the museum path introducing an polycarbonate back-lit exhibition guide that runs throughout the sequence of alternating high and narrow vaulted sequence. The exhibition system consists of a series of different elements designed according to their function and dimension: some are transparent show-cases, some other are continuous opaque lighted planes. All the exhibition areas can be easily arranged in different configurations through the adoption of a number of complementary display elements that can be added to the above mentioned linear system. These elements could be roughly divided in the following categories: display shelves, seats and information panels.

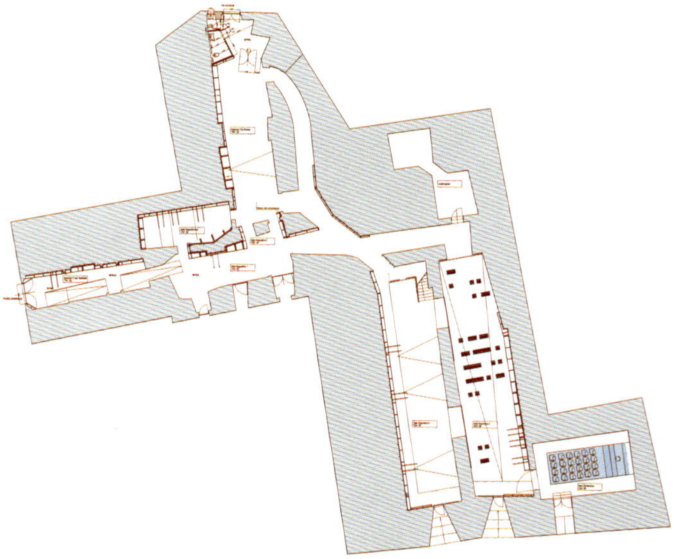

在Pitigliano的历史背景中，通过一种展示街的设计在Rocca Orsini的古老建筑中加入了一条城市通道。此设计旨在突出该博物馆通道的连续性，此通道贯穿了交替分布的高而窄的拱状系列展示区。此展示系统由一系列根据其功能和规格尺寸设计的不同成分构成：包括透明的展示柜、有灯光装饰的连续不透明平面。通过选择一些补充展示成分加入到上述线形系统中，可轻易将所有展示区安排成不同的配置形式。这些成分可粗略分为如下类别：展示架、座位和信息板。

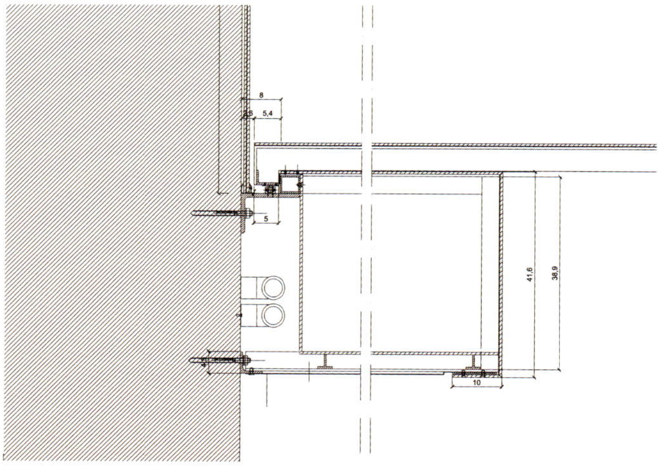

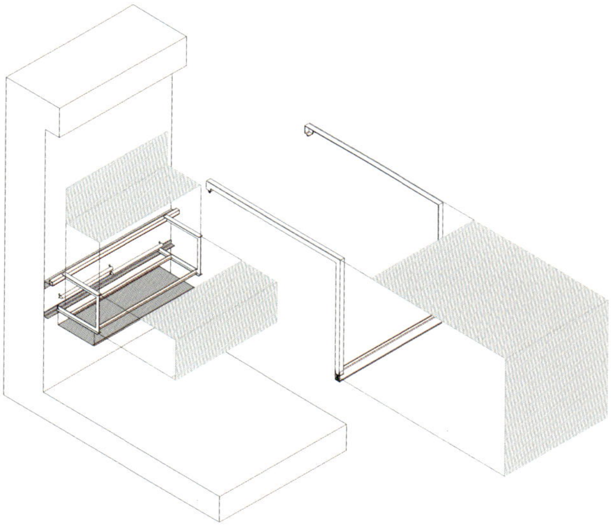

Architectures 4

bennet
CASARTUSI

Casa Artusi

Casa Artusi文化中心

Cultural Centre Casa Artusi, Forlimpopoli | Design: n!studio Susanna Ferrini, Antonello Stella | Team: Davide Gamba, Stefano Battaglia, Simone Bove, Gabriele Misso, Giulia Scaglietta | Consultants: Carolina De Camillis, Riccardo Fibbi Lighting Design, Pmg Group Budgets and Specifications, Enrico Bonafede Studio Volk with Stefano Battaglia Graphic Design | Client: Forlimpopoli Municipality (Fc) Contructor: Langella S.r.l. Napoli | Period: 20021-2007

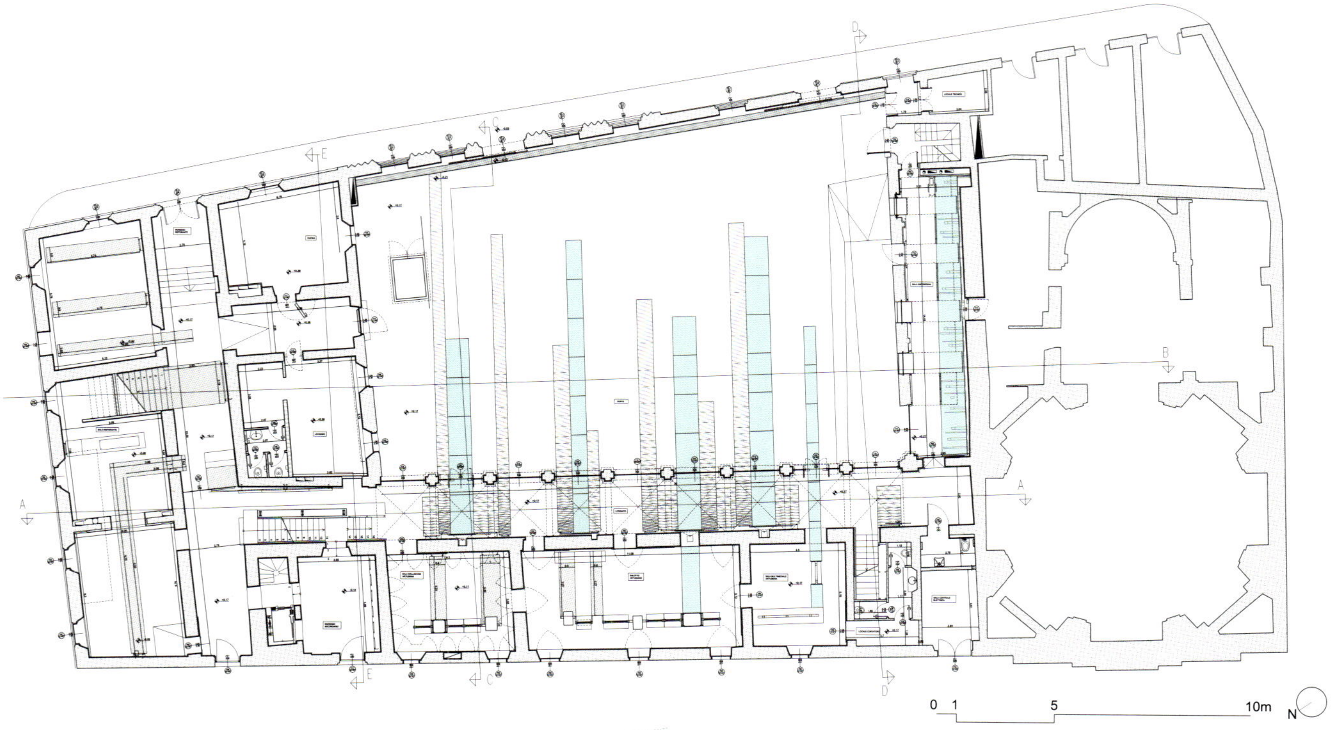

CASARTUSI

Casaartusi is the first centre of gastronomic culture for the promotion of home-cuisine, eternally bound with the name of Pellegrino Artusi, born in Forlimpopoli.Within the complex of the Chiesa dei Servi, an ancient cloister of unquestionable architectural and historical value, Casaartusi lies on a 2800 M^2 area and includes a peculiar mix of various facilities: permanent and temporary exhibition areas; specialized and generic library; one home-cuisine restaurant; spaces for tasting food and wine; conference room; cooking school etc. The core of the project is the historical entrance hall, a slightly declining plane designed with glass bands inviting visitors to walk through the porch, entrance of Casaartusi. The leading idea of the design is to interpret the different special and functional units of the historical complex as a whole entity, as far as materials and visual perception of the setting are concerned. The story of Mr. Artusi becomes the unifying element that ideally and visually connects the whole building. The insertion of alternating glass and steel stripes creates a single system that strongly characterizes the sense of the interior space. The intervention features 3 materials plus 1: steel, glass, polycarbonate and stone for the flooring, constantly arranged with small variations to better capture the sense of Artusi's story.

USI

CASARTUSI
LABIBLIOTECAINCUCINA
L'ARTE CULINARIA
LASCIENZAINCUCINA
L'ARTE DI MANGIAR BENE
ILRICETTARIO

LAGASTRONOMIA
IGIENEESALUTE
CASARTUSICASARTUSICASARTUSI

20
7
184

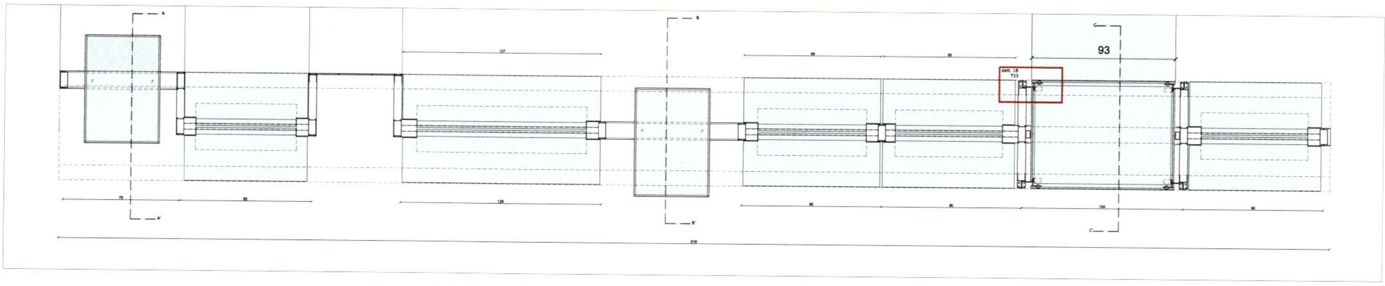

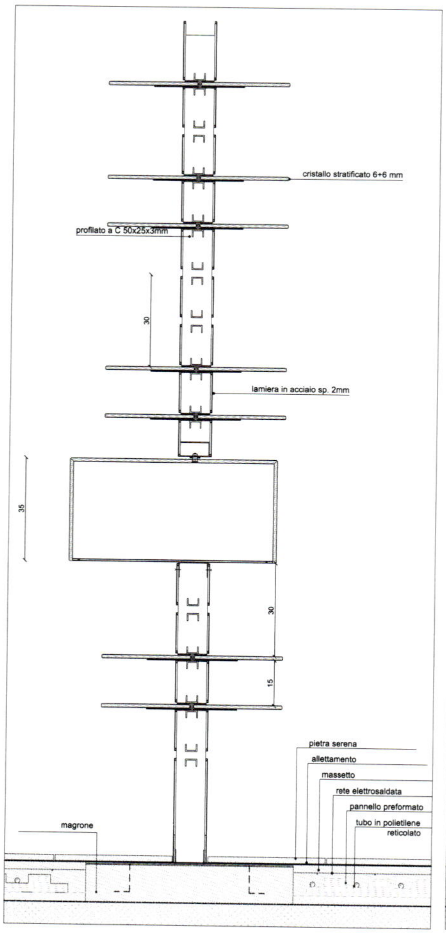
cristallo stratificato 6+6 mm
profilato a C 50x25x3mm
lamiera in acciaio sp. 2mm
pietra serena
allettamento
massetto
rete elettrosaldata
pannello preformato
tubo in polietilene
reticolato
magrone

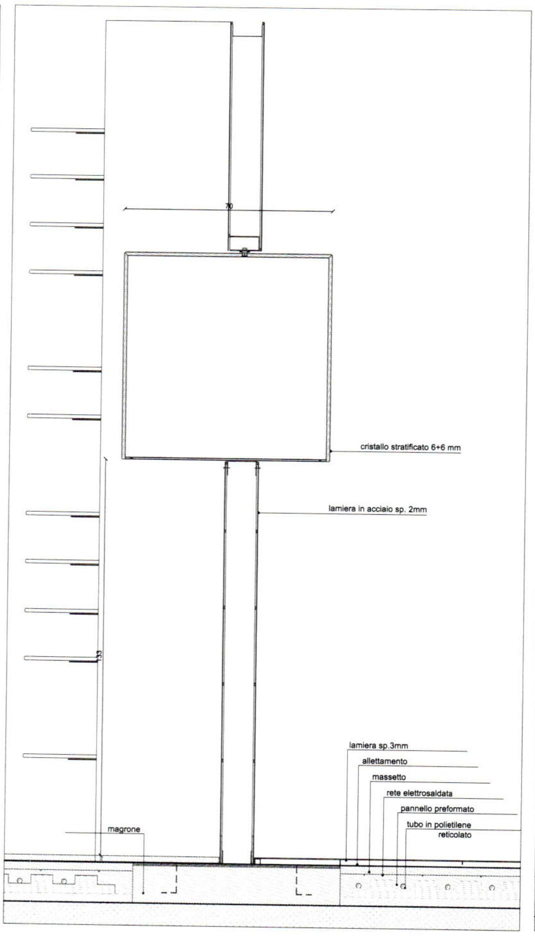
cristallo stratificato 6+6 mm
lamiera in acciaio sp. 2mm
lamiera sp.3mm
allettamento
massetto
rete elettrosaldata
pannello preformato
tubo in polietilene
reticolato
magrone

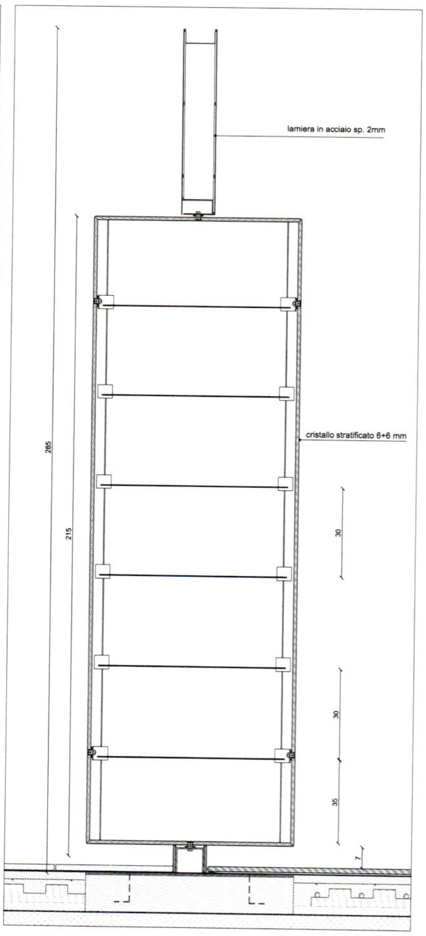
lamiera in acciaio sp. 2mm
cristallo stratificato 6+6 mm

ARTUSI
Biblioteca

20
47
209
galleria

CasaArtusi是第一家为推广家庭烹饪而建立的美食文化中心，该中心以Pellegrino Artusi命名，建立于Forlimpopoli。CasaArtusi位于Chiesa dei Servi修道院这座具有无可非议的建筑和历史价值的复合型建筑中，占地2800平方米，它是各种功能设施的组合体，包括：永久性和临时性展示区、专业和普通图书馆、一家家庭烹饪饭店、食物和红酒品尝区、会议室、烹饪学校等。该项目的核心是其历史悠久的门廊，其具有玻璃镶边设计并稍微倾斜的地面欢迎着游客们，即CasaArtusi的入口。此设计的主要构思是，在对这些设置的材质和视觉感知方面，将此历史性综合建筑的不同特色及功能部分诠释为一个整体实体。Artusi先生的经历成为了将整座建筑连为一体的理想视觉纽带。交替嵌入的玻璃和钢制条带形成了一个能够强烈体现其内部空间特色的单一系统。其设计的特色体现在材质上：地面采用钢、玻璃和聚碳酸酯加上石头制成，统一布置并有微小的变化以更好的体现Artusi的经历的意义。

Jean Prouvé, maison coque

House new motion

It is always better to observe the house
through the eyes of those who inhabit it
figures often move with evident dissonances
or they occupy aseptic containers through controlled processes of appropriation

by Susanna Ferrini

from PPC Magazine n. 20 21, University of Chieti 'G. D'Annunzio', Faculty of Architecture-Pescara

imagining the dwelling/a certain form of time

Only with great difficulty to terms such as movement and temporality manage to transmit a 'sense' of the future, and only when they are in harmony with the paradigmatic change of the 'subject': while it may appear to be too direct to transmigrate the meaning of exclusively architectural attributions (visualised in cells, containers, prefabricated modules, articulation/installation), the risk is progressively more that of giving them the dimension of an object, retinal and no longer fully relational; within our landscape/world, to quote Merleau Ponty, it is the meaning, and not the object of design that has become transitory, evasive and mobile, not yet form but relation. On the other hand, it is precisely the multiplicity of periods of dwelling, and the contemporary condition that has led to the dissolvence of consolidated concepts of dwelling into an equal number of processes/phenomena and spatial and material densifications and matrixes that are frequently more related to industrial processes than those of composition.

In our opinion we are no longer dealing only with a question of form, but rather a sense of objective reality and its transformation and new interventions. It is not only, or no longer, exclusively a question of form. The idea of dwelling appears to move faster than the transformations of its modalities, accelerating a process that dissolves the instruments of design, progressively more unsuitable to the era in which we live and its new methods of dwelling and the 'possibility' of its existence in multiple and different forms. The idea, understood as the meaning of design, appears to be longer lasting than its 'forms'. Only through its re-foundation can we imagine a future horizon of common research in an era of the hyper-proliferation of images such as ours. In the methods of the 'existence' of dwelling it is thus necessary and possible to re-trace a continuity of thought, without necessarily seeking a continuity of form.

The design of dwelling cannot help but accept a 'certain form of time', not as the problem of the 'from of objects', but as the rhythm and perception of the matter of which the process and the event are composed. Contemporary dwelling is

made denser by the multiplicity of rivulets in which we lose the individuality of subjects, or the new networks of un-programmed communities or the separations generated by religious, ethnic and behavioural foundations that are grafted onto the physical or virtual landscape; changes in paradigms resulting from changes in the 'position' of the subject, of our observation.

The house becomes a tool of perception, another gauge for measuring transformations, and one more sense that is useful to our understanding.

Michel Serres used the following words to describe our existence, pervaded by the interweaving of the five senses that, when mixed together, define our body through the experience of inhabiting an infinite number of boxes, be they hard or soft, individual or collective, as if they we our skin, our houses, our cities: "The box, useful for understanding, is necessary to our existence. I am it. I inhabit it". The meaning of inhabiting shifts from the conception of a space/place of localisation to one of the subject/existence. The architectural meaning appears to wander, following our physical re-positioning in space and the trajectories that I use to define my dwelling as a total existence, as the network that defines successive circuits and 'boxes'. We are the subjects who, through the act of dwelling, discover new spaces, such as, for example, the spaces of infrastructure, abandoned structures and the new spaces of work.

house new motion

As part of this reappropriation of the landscape/world, the succession of interventions in casa new motion represents a reflection on new ways of dwelling based on a private dimension and a more collective one, whether they tend to be typologically integrated with the spaces of work and recreation, or whether they contribute to the very form of the urban and diffuse landscape. The space of dwelling 'crosses' the new methods of design, construction and the new 'significance' of the spaces of public life, appropriating the tensions of contemporary thought, such as flexibility and the possibility to transform spaces and structures, categories that strongly introduce the dimension of time within that of space.

The reality of dwelling is rich with different scales of design, from the minimal unit to the cell-cockpit-prototype, progressively reaching the most stratified articulations of spaces and functions, the point of the interweaving of networks, infrastructures and urban systems, as interpreted by different temporal dimensions, from transitory inhabitation to the variation and growth of nuclei and aggregations of families.

Technical experimentation with the very 'matter' of construction, as part of a new case-study of components, materials, building systems and installations is involved, as a figure in the foreground, with the ability to activate articulations of space and movement, using design to invert the primacy of form as the complete and compositionally 'finite' dimension of construction, focusing instead on versatile processes of industrial technique and different methods of user appropriation. The importance of this transformation lies in a new tension and tectonic re-founding of the process of building.

Is it possible to speak of beautiful and ugly works of architecture, or only those with and without 'meaning'? The 'meaning' of the design process guides and marks the process of building and provides the genetic code of education, together with that of the changes and transformations that exist in the realm of its potential. We must move away from the standard collection of designer homes for excellent clients and rediscover the unsettling and transversal dimension of the everyday, of the common event, of the anonymous and widespread movement of building the city and the landscape.

Dwelling is thus intended as the most complete dimension of 'orientation' within and 'caring' for space; this makes it possible to 'imagine' multiple ways of dwelling through design and to use the precise shifting of functions and non-coincidence to 'define' the threshold between different environments and temporal experiences.

Within the dimension of the process/event, each project becomes an ethical-existential process that wraps the evolution of an idea: a continuous 'search for meaning'.

In similar terms, man's perception of the external world, his ability to amplify it or modify it in light of new developments in the field of science and philosophy, cannot help but stimulate a continuous and intense process of spatial 'appropriation'.

Through this form of action, the reverberation of the sign in the landscape/world that surrounds us appears to be more important than the sign itself, and thus the process of transformation that it leads to and undergoes during its successive phases is given more importance than its fixed and definitive form.

Thorough his inhabitation of the modern metropolis, modern man came to understand the possibility of existing only in a crowd, one individual amongst a multitude of individuals.

The 'possible' solitude that René Descartes theorised in life in Paris, in an image that is a sharp rebuke, reveals, for the first time, the 'possibility' and the choice of being alone in a crowd, in a plurality of images and transformations. The horizon of solitude in the metropolis indicates not only our desire to disappear in the crowd, but above all to 'choose' our existence and position within a continuous system of relations, in which we feel that we can contribute, through our bodily existence, to the general entropy of both being present and disappearing. Modern man thus chooses to live these 'differences': the dissolution of unity revealed by modern thinking in a pervasive realism that records the incessant transformation of reality and identifies beauty in the fleeting and the transitory, in change.

the distance that separates us from the world

How does the contemporary dwelling represent our

observation and idea of inhabiting the landscape/world? More and more we favour less and less: less matter, less form and less function appear to focus us in an ontological and paradigmatic way towards a desire to diminish the threshold and erase the distance that separates us from the world. All the same, we move less and less towards the maximum level of transformability and variety, towards the definition of a new idea of the architecture/sustainability of technical and functional services. The shift in functional meaning, no longer only the home, but rather a structure inclusive of functions and meanings, no longer only an internal space, but a spatial and perceptive continuity with the ground-landscape leads us towards estrangement and the cancellation of the limit and the threshold. We attempt to overcome formal and conceptual categories and envelop new ones, in order to cancel the distance between our presence/existence and the objective reality that we live in, that contains and guides our behaviour and our very existence, with respect to which we feel discomfort about a non-adherence with our existence, our desire for appropriation and recognisability. This is the duality, which has always existed by definition, between content and container, between that which wraps and protects man, this series of boites, skins, envelopes, to use only a few terms, which now seem to explode as a result of their unsuitability and incapacity to meet our needs and expectations. On the other hand, they remain inanimate boxes that do not represent us and whose walls reflect the dull thuds of our tensions with the exterior world. It is precisely the desire to cancel the distance between ourselves and the world and the diaphragms that obscure our view of it that highlights the new foreground condition of the subject as part of an amplified concept of functionality that overruns the very structure of architecture, disappearing in a self-referential and disciplinary form, in order to include perception and individual and collective sensibility amongst its genetic baggage. This condition led to the period of investigations of envelopes that were deformed to match the subject, a period that can be interpreted as the manifesto of a primordial formal coincidence with the being and the subject, malleable to temporal and local transformations.
THE MINIMUM MATERIAL POSSIBLE as the maximum cancellation of the perception of matter. THE MINIMUM DISTANCE BETWEEN OURSELVES AND THE WORLD as the idea of cancelling, as much as possible, the distance between the observer and the object. THE DISTANCE BETWEEN REALITY AND ITS PERCEPTION, to the point of negating design if it becomes the separation between ourselves and the world, most likely resulting in individual and extreme forms of appropriation and the individual and multiform use of space, based on the use of a list of components.
Research into dwelling appears to generate and develop precisely in the simultaneity of sharply differentiated scales, almost as if it were guided by the opposite ends of a generative process that is left to its uncontrolled 'growth': a general vision of phenomena and, at the same time, a vision under the microscope, an up-close analysis of the molecule-cell-unit, fragment, generating component, DNA, capable, through its genetic 'prefiguration', of controlling the future configurations of dwelling.
The catalogue of Fuller's "4D Time Lock", his first elaboration of the Dymaxion House published in 1929, featured a dense theoretical and programmatic definition, reiterating the very aspects of 'imagining' architecture as a prototypical prefiguration, in which the 'description' of the elements, their nucleation as a molecule that can be traced back to everything, to the presentation of the construction site-assembly, becomes more important than the real object and its form. In a separate chapter this is how he describes the passage, only apparently technical in form, of the building "From the Inside Out" to its opposite "From the Outside In". The generative process that moves from the Interior towards the Exterior thus becomes a privileged view of an explosive and multiplicative process of the subject, of its continual articulation in space. This results in the progressive confirmation of the equation Imagining Design-Imagining the Dwelling. The dwelling, more than being an objective reality, becomes something with the inherent quality of an event. This tension works towards the definition of a 'sensible environment', in which our perception can be modified through interaction with the environment: a focus on the body features an awareness of the behaviour of forms of spatial appropriation, broadening the field of architecture. This is a modality of the existence of a subject that exists at the 'limit' of dwelling, in a continuous oscillation between flexibility and appropriation. A condition that signifies thinking more about the event than the object. The limit is the space in which the exterior is also interior, where it is possible, from the interior, to 'lean out' into the exterior. The basic reduction of any representation of function leads back to the minimum use of energy and the search for the maximum level of complexity. The history of constructive thinking in

the field of prefabrication pre-announced the current state of development. Questions and inquiries were raised about the existence of design, focused on its contraction into primary acts, precisely when we move in 'another' dimension with respect to the world of representation.
The idea of design thus involves not only the methodologies, but also the very physical consistency of the architectural object. The dwelling becomes the prototype, the space of successive significations tied to use, which tends to develop and adapt to different periods of time and different spaces. The space of dwelling is rendered 'neutral' and can be interchanged with different functions. It is home to the fusing of the 'limit' between the permanent and the temporary. The appropriation and transformation of the individual becomes a fundamental theme that presupposes the 'prefiguration' of possible future transformations. Dwelling as an in-situ event of the object reproposes the differences that remain in the praxis of the 'limit' of contemporary design, such as the dualities house/space/prototype, house/body/virtual and new urban alternatives, the vast and inclusive meaning of the social act of inhabiting the variability/multiplicities of urban, ethnic and religious conditions.
We can refer to a comment made by Studio Azzurro, "paradoxically, the more things tend towards materialization, dissolving their physical qualities in flows and data and tending to disappear, the more their cumbersome presence emerges". Never as now has the virtual become a cumbersome presence, almost a redundant and reverberating physical quality in the hyper-proliferation of a world of forms. The idea of dwelling can no longer be a reflection of the simple act of inhabiting the traces of invisible corporeality, imagined as an abstraction, as a possible reflection on the limit. In a plurality of images, domesticity appears to be confirmed only by its disappearance, once again taking advantage of abstraction

in the same way that we can reflect on the concept of the 'limit' through its negation, re-founded in its appropriation of new aspects of physicality and corporeality, which is once again like a change and transformation in the paradigm of design.

住宅新潮流

观察房屋建筑的最好方法
就是通过其居住者的眼睛来进行观察
数字通常会随着明显的不协调情况而变化
或他们会通过受控制的分配流程占据洁净的容器

Susanna Ferrini

1 设想居住模式/一种特定形式的时间

只有以很困难的形式，且只有在与“对象”的典型变化情况相协调一致的情况下，如运动和暂时性等术语才能够传达出一种未来的“感觉”。而仅仅转述建筑属性（体现在房间、容器、预制安装模块、联接/安装等方面）似乎又太直接了，其所涉及的风险越来越多地集中在赋予某件事物维度属性方面，对于视觉方面而不再完全关联；在我们的世界中，引用Merleau Ponty的话来说，是意义，而不是所设计的事物，已经变得短暂、不可捉摸且可动易变，还不是形式，而只是关系。另一方面，工业工艺在更多情况下正是与居住阶段的多样性，以及将固定的居住概念终结而发展为同等数量的过程或现象的当代条件和空间和物质浓缩和基体相关，而不是与那些成分相关。

我们认为，我们所面临的不再完全是一个形式问题，而是客观现实的感觉以及它的转变和新介入模式。它不仅，或不再，完全是一个形式的问题。住宅概念的发展似乎比其形态的转变速度要更快一些，加速了使设计工具失去作用的过程，它越来越不适合我们所处的时代、其住宅新方法及其以多种不同形式存在的“可能性”。概念，我们可以理解为设计的意义，似乎比其“形式”的寿命更长久。在如我们所处的这个影像技术超速膨胀的时代，我们只能通过重建的方式才能够想象到普通研究的未来水平。这样我们才能够在不必追求其形式的连续性的情况下，通过住宅“存在”的方法重新追溯思维的连续性。

住宅的设计不得不接受一种“特定形式的时间”，不是以“物品形式”问题的形式，而是作为过程和事件的组成事物的节律和感知。我们在川流纵横的繁琐中丧失了物品的个性，导致当代的住宅更加密集，或导致建造出了未经规划的社区新网络，或将宗教、种族和行为模式基础转植到了物理或实质性风景中形成了各种隔离区，因物品的位置以及我们的观察效果的变化而导致了典型模式的转变。

房屋变成了一种感知的工具，即另一种测量转变情况的标尺，它具有更多对于我们的理解更为有用的感觉。

Michel Serres用下面的一段话描述出了我们的存在，充满了相互交织的五种感觉，当这些感觉相互交融在一起时，就可以通过占据无限的盒子的经历来定义我们的身体，无论这些盒子是硬还是软，是单独的还是集成在一起，就好像它们就是我们的皮肤，我们的房屋，我们的城市：“这些盒子有助于我们的理解，它们是我们的存在所必需的。我就是它。我住在其中。”

居住的含义从定位的空间/位置的概念转变成为了物品/存在的概念。随着我们将住宅定义为一个整体存在时所用的空间和轨迹方面的重新定位，就像将连续的圆圈定义为“盒子”的网络一样，建筑的意义似乎在发生着漂移。我们就是通过居住行为发现如基础设施空间、遗弃建筑和新建筑物空间等新空间的人。

2 住宅新潮流

作为此项对地貌风景或世界重新进行分配行为的一部分，casa新运动的系列介入活动反映了以私有空间和更为集体式的形式为基础的新型居住模式，无论是否倾向于从类型上与工作和娱乐空间集成到一起，无论是否有助于形成城市和分散式地貌风景形式本身。居住的空间“跨越”此类设计的新方法、建筑和公共生活空间的新“意义”，对当代思想的压力进行了分配，这些压力包括转变空间和结构的灵活性和可能性，将时间的维度引入空间的维度的客观实物。

在住宅的现实情况中，设计范围多种多样，从最小单元到座舱驾驶舱原型，逐渐发展到最为层次化的空间和功能的联接，达到相互交织的网络、基础设施和城市系统的一点。就像各种时间维度所阐述的一样，从短暂的居住到核心的变化和增长，以及家庭的聚集。

就建筑的“事物”本身所进行的技术试验，作为一项针对部件、材料、建筑系统和安装所进行的新案例研究的一部分，是最为显著的数据，可以激活空间和运动之间的联接，通过设计将形式的私密性转化为建筑完全、综合的“有限”纬度，转而集中在工业技术的通用流程和用户应用的不同方法。此转化的重要性在于一种新的压力和建筑物流程的构造重建。

是否无法论及建筑作品的美或丑，而只能论及其是否有意义？设计过程的“意义”为建筑过程提供了指南和标记，提供了教育的遗传编码，以及其可能领域的变化和转变。

对于极好的客户我们需要撇开那些标准的设计工作室，重新研究那些日常工作中尚未解决的横向维度，那些常见事件，那些建造城市和风景园林过程中未知的、普遍的活动。

因而住宅就被当作了空间“定位”和空间“护理”方面最为完整的维度，这样我们就可以通过设计来“设想”多种住宅形式，并利用其准确的漂移功能以及不符之处来“界定”各种环境和短时性经历之间的临界线。

在过程或事件的维度，每一个项目都变成了一种围绕某一构思的演变而进行的道德存续过程：一种持续“寻求意义”的过程。在类似的情况下，人们对外部世界的感知，在科学和哲学领域出现新发展时将其放大或进行变更的能力，将不得不存进一种持续、高强度的空间“分配”程序。

通过这种形式的行动，围绕在我们周围的风景或世界的标志的反射看起来似乎比标志本身更重要，因而其随后阶段过程中所导致和经历的转变过程比其固定和最终形式更为重要。

通过居住在现代大都市，现代人了解到，他可能只能在人群中存在，单个个人生活在多个个人之中。

Rene Descartes以一种强烈指责的形象，在巴黎生活中以理论形式总结出的“可能”独居的形式，第一次揭示了在人群中，在多种形象和转化中保持独立的”可能性”以及对这种方式的选择。在大都市中独居的见识不仅揭示了我们对于消失在人群中的渴望，最重要的是要“选择”我们存在的方式以及我们在一种关系的持续系统中的位置，在此系统中我们感觉能够通过我们的存在来影响存在和消失的整体平均信息量。因此现代人选择要经历这些“不同之处”：记录了现实的不断转变并确定了飞逝和短暂在变化中的美感的现代思维模式，通过一种普遍深入的现实主义所揭示的联合形式的解体。

3 将我们同世界隔开的距离

当代住宅如何能够体现我们对在世界中居住的观察和看法？
我们越来越倾向于越来越少：更少的事物、更少的形式以及更少的功能，以存在论和典范的模式使我们集中考虑清除临界线以及去除将我们同世界隔开的距离的愿望。同样，我们越来越少地提倡最大程度的可变换性和多样性，或对技术和功能性服务建筑以及对可持续发展新思路的定义。它不再仅仅是家的概念，而是一座包括功能和意义的建筑结构，不再仅仅是内部空间，而是与地面风景在空间和感知方面的延续，这种功能性意义的转变使我们远离并取消了这种界限以及临界线。我们试图克服正式和概念性的类别并包罗入新的类别，以清除我们的存在同我们所居住的，容纳并指导我们的行为和我们的存续，在这方面如果不忠于我们的存续我们便会感到不适，我们对分配和认可的期望以及与客观现实之间的距离。这就是其双重性，正如其定义所述它一直存在于内容及其容器之间，在限制和保护人类之间，此系列包括餐馆、表层、包层等，因为它们不具有可持续发展性且不能够满足我们的需要和期望，现在看来它们似乎即将被废除了。另一方面，它们保留了不能够代表我们的死气沉沉的盒子，它们的墙体反映出了外部世界间的压力的枯燥重击声。
正是我们想要消除我们同世界之间的距离，以及使我们对其看法模糊不清的隔阂的期望，着重凸显了此物品作为超出了建筑结构本身被放大了功能性概念一部分的最显著的条件，以自我参考和训练有素的形式消失，来在其遗传包中增加感知和个体及集体感受能力。因为这个条件使我们花费了一定时间对被改变形式以适应物品的包裹物进行了研究，此阶段可以被解释为存在和物品最初正式重合的宣言，很容易适应时间和地点的转变。
最大程度事物感知消除的最小物质的可能性。我们同世界之间的最小距离就是尽可能消除观察者和被观察物品之间的距离。现实与其感知之间的距离，在如果它成为了我们自己和这个世界之间的隔离之后便需要放弃该设计的程度，最可能导致单项、极端形式的分配对空间的单独使用以及多形式应用，根据所使用的部件清单而定。

对住宅进行的研究似乎能够准确同时生成并发展出鲜明不同的规模，几乎就像是在其不受控制的“生长”过程所使用的正好相反的生成过程的指导下进行的一样。现象的一种总体景象，同时，一种显微镜下的景象，对分子单元、片段、生成元件、DNA的密切分析，能够通过其遗传“原型”控制未来住宅的配置。
Fuller的“4D时钟”一览表，即他于1929年发表的第一份对Dymax-ion House的详细论述，重点论述了一种浓厚的理论和程序性定义，重新阐明了“造影”建筑作为一种原型配置的各个确切方面。其中对自然环境的“描述”，将其集结成一个可以追溯到任何事物的分子，用于说明建筑现场的情况，这比其真实物品和形式更为重要。他以同样的方式在另一个章节，只是明显是以技术的形式，以“从外到里”的模式而不是“从里到外”的模式对建筑进行了阐述。从内部向外部移动的生成过程因而就成为了该物品的激进型和倍增型特许视角，其在空间方面的连续联接。

这就导致逐渐确认了想象设计–想象住宅方程。住宅不仅仅是客观存在的，它成为了一种具有事件的固有性质的事物。这种压力向一种“敏感环境”的定义努力，其中我们可以通过与环境的相互关系对我们的感知进行变更：重点放在身体上表明已经注意到了空间分配形式的行为的重要性，扩展了建筑的领域。这是存在于住宅“范围”内的物品以在灵活性和分配之间摆动的形式的形式。这种情况更注重事件而不是物品。其范围就是外部同样也是内部的空间，在可能的情况下，从内部来“向外部倾斜”直到外部。

任何功能表现形式的基本降低情况都会返回到最低的能量使用量，以及寻找最大程度的复杂性方面。
在预制制造领域的建设性思维历史性的预先宣告了当前的发展状态。针对设计的存续方面提出了一些问题和质疑，主要集中压缩到主要行动方面，确切地说就是我们在表征世界移动到“另一”纬度的时候。
因而设计的构思就不仅仅涉及到方法学，还涉及到建筑物品的物理持续性本身。
住宅变成了一种原型，捆绑在一起使用的意义的空间，它倾向于不断发展并适应各种不同的阶段和不同的空间。住宅的空间被定为“中性”，可同各种不同的功能互换。它是融合永久性和暂时性之间的“界限”的基地。个体转化的分配成为了预示未来可能出现的转变的“原型”的基本主题。
住宅作为物品的现场事件再次提出了存在于当代设计的“界限”惯例中的差异，例如房屋/空间/原型，房屋/主体/真实的新城市选项的二元性，居住于城市、种族和宗教条件这种可变性/多样性情况下，这一社会活动的巨大和包罗广泛的意义。

我们可以引用Studio Azzurro所作的评论，“自相矛盾地来说，事物超物质化发展，将其物理特性分解成将要消失的流程和数据的趋势越多，就会出现越多的棘手事物。”实质性从来没有像现在这样成为了一种棘手的事物，几乎是这个形式的世界中过度扩散的一种多余、混响的物理质量。住宅的概念不再是占据隐形存在的痕迹这一简单行为的一种反映，想象为一种抽象概念，一种对界限的可能反映。在多种形象中，似乎只能通过消失来证明家庭生活的存在，这又一次利用了抽象的概念，就像我们能够通过否定“界限”来反映其概念一样，在其物质和形式的存在方面重新建立了其分配情况，同样就跟设计范例的变化和转变一样。

Museum as an icon of urban space

by Susanna Ferrini
from WA-World Architecture n.195 2006, Tsinghua University, Beijing

In the field of the contemporary architectural research, the museum has assumed a central role in the configuration of the city and public space.
If it is true that two museums are inaugurated around the globe every week, we can thus confirm that the space dedicated to cultural communication is increasingly more often the symbol of an awareness that is extended to broader sections of the population.
In recent years, it is precisely the construction of new museums that has highlighted the central role of architecture in the processes of transforming the city, establishing what can almost be called the primacy of architecture over urban planning.
This has strongly influenced architectural research: not by chance the two most celebrated buildings by official culture from the last decade, the Guggenheim Museum by Frank Gehry in Bilbao and the Tate Modern in London by Herzog & De Meuron represent, even in the diversity of their interpretations, the central role of the museum building as the driving theme of architectural research. The resonance obtained by these buildings, above all beyond specialised research, convinces us of the fact that the space of the museum is becoming the privileged space of cultural encounter and communication for the media and the general public.
Thus the museum increasingly presents itself as the space of public encounter and the expression of new cultural horizons. The museum tends more to 'represent', in a symbolic manner, the event of this public rite rather than limiting itself to the simple display of works of art.
In parallel there has also been an evolution in the approach of the visitor/user of the contemporary museum, leading to a differentiation of the levels of use of the museum: from a first level characterised by a rapid and basic understanding aimed at collecting only a few key concepts about the topic, to the request for more profound levels of awareness tied to the diffusion of an increased level of interest in artistic events and an increased central role of free time and cultural tourism for the masses.
More frequently the public/visitor becomes the active subject of architectural form. His/her visual and sensorial perceptions are stimulated through spatial articulation and the choice of materials. In fact, the two buildings mentioned above, like the majority of contemporary museums, are characterised precisely by a large interior space that serves as the entrance: at the Tate Modern the void of the large turbine hall of the Bankside power plant is roughly one third of the overall volume of the building and the largest public space inside a building in the world; at the Guggenheim, the central atrium, with a surface area of 300 m^2 and a height of 50 metres, has no direct function as a display space, but rather represents a new public space for the city, with the objective of creating an absolute continuity of flows and paths between the interior and the exterior of the building.
In fact, a theme that has been widely dealt with in modern and contemporary debate is the search for an equilibrium between the container and the content of the museum, between the architectural form of the envelope seen as a collection of surfaces for communication and the content of what is on display.
Museums have become large icons of contemporary urban space, with a prevalence of the figure of the 'container' as the condenser of flows, events and information and thus a recognisable sign within the city. The envelope of museum buildings is called upon to provide a growing level of intensity, emotion and perception, coupled with the achievement of new levels of interaction between the visitor-user and the simple and immediate communication of content.
Museum buildings have managed, in many cases, to catalyse urban space through their force and determine its transformation. Furthermore, the progressive lack of trust in urban planning and the idea of the plan as the ordering element of urban phenomena has led to architectural interventions with highly symbolic meanings, which are also bearers of a strong urban identity.
Without a doubt, since the 1980's we have been witness to a

transformation of the logic of intervening in the city; the objective of the new large museum projects is precisely that of affecting and directing urban transformations. A trend that has marked the end of the image of the museum as a closed and self-referential 'box', a 'backdrop' for the conservation of artistic treasures that have no relationship with their exterior environment. Fundamental projects of this type include the Kunsthal by Rem Koolhaas, Steven Holl's Kiasma in Helsinki and the Institut du Monde Arabe in Paris by Jean Nouvel, where geometry and the urban fabric are reflected in the very form of the building.

Today we ask ourselves about the reasoning that has led the typology of the museum to assume such a central role in architectural research and more in general to represent contemporary cultural evolution. In fact, there appears to be no end to the proliferation of museums throughout the world and in Europe, leading Harald Szeemaan to state "for an architect today building a museum is akin to what is was once like to build a cathedral", generating the paradoxical condition that "there is one museum for every 9,000 inhabitants in Switzerland".

On the one hand, a positive aspect is that art is no longer closed inside a self-referential elitism and that architecture seeks a re-founding of the discipline in the vast sharing of its objectives and in the achievement of a greater degree of 'empathy' with the visitor. On the other hand, the danger is that this tendency leads to an excessive spectacularisation of the architecture of the museum, with the attribution of an excessive meaning to architectural form with respect to the content being displayed; a prevalence of the desire to 'surprise' over the 'sense' of the theme of the museum.

In many cases the form of the museum interprets the theme, becoming the work of art to be displayed; not by chance the Judisches Museum in Berlin, designed by Daniel Libeskind was visited, even before the exhibitions were set up, by over 350,000 people and it should not surprise us that the internal exhibitions seem to be 'superimposed' respect to the perception of the architectural form and the sensations that it intends to transmit.

One possible consequence of this conception of the envelope as museum-container: a large 'box' that is formally defined on the exterior, but 'neutral' on the interior in which the museum display is designed in a second phase, overlapped indifferently like a second 'skin' on the interior.

Significantly different was the lesson of the masters of the 'Modern' movement in the conception of a new space for the museum. In the work of architects like Frank Lloyd Wright, Louis Kahn and, in particular in Europe, Carlo Scarpa and Franco Albini, whose respective Castelvecchio Museum in Verona and 'Museo del Tesoro San Lorenzo' in Genoa bear some mention here, the concept of the exhibition was born in unison with that of the building. In these projects, the architectural theme and that of the exhibition appear as an inseparable 'whole'; in particular, the central theme being dealt with is the relationship between the user-visitor and the content of the exhibition, a process within which the architect becomes a medium, a guide to understanding, though never a diaphragm or an obstacle.

The museum as an icon and sign of strong formal and visual intensity highlights the existence of a debate about the dichotomy between the museum as a 'neutral' space and the museum as a 'work of art' as advanced by Frank Gehry in an essay in 1998. The theme of the 'neutrality' of the museum in relationship to Gehry's work appears to be an unrealisable myth: "Artists prefer to display in a strong building, they do not want to show their work in a neutral space". Even if the conceptual approach differs, Jacques Herzog also underlined the importance that architecture can affirm itself as architecture and not as a negation of itself: "I think there enough museums now with simply designed galleries satisfying the need for plainness "(...) The Tate Gallery, in the former Bankside power station, will certainly have essentially peaceful galleries. But there will also be a dramatic quality. We want to stretch the senses between large and small spaces."

Regarding these aspects, even in contemporary design, it is hoped that we can find a thread of continuity with the research of the masters who laid the foundations for the 'modern' conception of the museum. The objective may

be that of a renewed synergy between architecture and the exhibition, in order to render the museum a flexible structure that is capable of transforming and evolving in its content. The image of the contemporary museum is thus that of a museal mechanism, technologically complex and advanced and with ample margins of flexibility in the articulation of spaces and functions, capable of renewing its contents and updating the multimedia systems of communication, in synthesis to follow and confirm the growing rhythm of transformations of the historical and scientific research.
New interpretive categories characterise the architectural research related to the contemporary museum in Europe and can be synthesised in a few interpretive concepts:

the museum as a part of the city

If we have to find an archetype of the urban conception of the museum it would undoubtedly be the Altes Museum designed by Karl Friedrich Schinkel on the island in the Sprea River in Berlin, between 1824 and 1830. The Altes Museum represents the initial nucleus of the first public museum centre in Germany and one of the first public buildings designed as a museum form its very inception. The creation of the island of museums continued for approximately a century, with a series of related buildings and based on a plan, including the Neues Museum, the Alte Nationalgalerie, the Bodemuseum and the Pergamonmuseum, built to display the large altar to Zeus found near the Pergamon hill in Greece.
The Museumsinsel, currently being completely renovated, presents itself as a complex museum-city system that is capable of strongly characterising the urban environment, an aspect that can be found in the large intervention in Vienna in the Museumquartet by Ortner&Ortner.
Thus, the museum-city is an urban infrastructure, the site of transit and urban flows. On the other hand, in Europe, cultural renewal is highlighted in the relationship with the design of large historical cities; the surgical insertion of the 'new' within the historical urban fabric has been possible above all through the construction of new museums that, since their inception, have been conceived of as symbols and icons of the city itself. The new European museum centres have been the object of transformation or ongoing changes to meet the new requirements of cultural tourism that continues to be more selective and diversified: Pei's glass pyramid at the Louvre in Paris and the new addition for the Islamic art wing by Rudy Ricciotti; the Uffizi confronted the important theme of a large entrance canopy designed by Arata Isozaki; the Tate Modern is already planning another addition to be designed by Herzog & De Meuron.
The strategy being implemented by recent museum projects is precisely that of guiding urban transformation.
Paris, London, Berlin, Vienna, Bilbao and Rome are only some of the most important cities in Europe that, during recent decades, have decided to renew their image through projects for and additions to large museums. The museum appears to be the ideal theme for manifesting the political and cultural life of a country and for promoting its identity abroad.
The emblematic case at the end of the 1990's was the construction of the Guggenheim in Bilbao, which managed to set off the economic and cultural development of an entire region, so much so that it is possible to speak of the "Bilbao Effect" as a synonym of development.
What are the expectations today for a new museum?
In fact, it appears to be impossible to speak of a singular idea or concept of the museum.
The institution of Museums and centres for the divulgation of research in contemporary art appears to be increasingly tied to general cultural phenomena, to a radical mutation in the approach of the general public to sites for art and culture. The museum is approached with a new sense of belonging to a global community that is part of a shared network of communication.
However, the museum is also the site of consumption, the consumption of the cultural image, the passkey to the acquisition of a code of shared communication, of an awareness that is often basic and hurried, where we wish to participate as representatives of the global community: an aspiration that architectural choices must necessarily deal with and for which it must provide answers.

multitasking

The experience and, if we wish, the effect of multitasking is found, using a metaphor of the digital, in the concept of the museum as the organisation of a complex series of spaces and functions. The museum becomes a mechanism that interprets one of the characteristics of the information society that can be identified in the possibility of living in synergy with the different experiences of contracted time with an average level

of attention. Multitasking thus deals with a new type of experience that is tied to contemporary society.
In more general terms the museum has undoubtedly been transformed into a complex system, museum and archive, a space for research and scientific divulgation, thus requiring multitasking at the level of experiences being proposed. The progenitor is undoubtedly the Centre George Pompidou, the machine that allows for the 'co-existence' of temporary display spaces for 'events', the rooms that house the collection, the large library and archives, multimedia spaces, conference and projection rooms and the section of shops and rest areas, including book shops, restaurants and cafeterias.

the global network of museums

The new dynamics and diffusion of mass cultural tourism have accelerated and produced new museum projects not only in large metropolis, but also in the smaller centres of art or above all in those urban areas undergoing transformation and in search of a new identity.
When delineating the new prospects of a contemporary society dominated by media and the information network, Paul Virilio speaks of the passage from a mass society, a phenomenon that spread during the second half of the last century, to one of mass individualism, characterised by a gloabalised behaviour and spread in a capillary manner by the network, even if tied to individual perceptions and experiences.
A fascinating aspect of this research is that the museum is a public space, but also the site of an individual experience of the subject-visitor, the site in which to capture a personal space of relationship with the works of art on display.
The concept of the museum in Europe has had to deal with the impact of mass culture: as part of this trend there is a further differentiation of the requests made by the museum to diversify the cultural offering and to further specialise the nature of its content.
What is being defined as part of the prospect for future development is the diffusion of a network of museums, where the Guggenheim represents the emblematic example with is construction of various museums around the globe: New York, Venice, Las Vegas, Salzburg, Berlin, Bilbao ...
"The concept of the Guggenheim is that of the international museum, spread out ever several countries, with discontiguous exhibition spaces. Our critics call our various museums 'franchising', or refer to them as a 'McDonalds' chain', but that entirely misses the point. The model of a 'franchise' implies total standardisation and a utter disregard for local context; the Guggenheim, in expanding its network of museums, takes the opposite approach. There are no two situations that resemble one another or in the slightest way are generically alike" Thomas Kerns says. These words highlight the new strategies of intervention: museums located in various countries, whose buildings are strongly characterised by context and theme. A model of a museum network that will increase the cultural 'nomadism' typical of current society.

the museum as envelope

The importance assumed by the museum in cities and the role of being an important cultural 'attractor' has certainly privileged the envelope in the perception of the building. Often, the design of the envelope wishes to underline the perceptive role of the building within the urban context, presenting a high degree of materiality, arriving at the paradox that amongst the works on display one can include the building envelope, its capacity to surprise and amaze. A trend towards the 'spectacularisation' of museum architecture that is highlighted both in the experimentation with new forms and the use of new materials, as well as in the search for the 'large' scale. Already, in the formulation of many programmes, the museum centre is presented as a 'citadel' that is self-sufficient in terms of scale and completeness and for the diversification of what it offers to culture, research and tourism.
Nonetheless, the attention focused on the theme of the building envelope is configured as the interesting search for a new materiality of the external 'skin'. Often the envelope is interpreted as an 'interface' between the museum and the urban space, becoming a system of communication for its contents. In the project by Herzog & De Meuron for the Museum of the Art of the XX Century, the exterior surfaces were characterised by digital images, representing the very idea of the evolution over time of the future contents of the

museum.
The search for a strong degree of materiality and meaning can be observed in the museum by Peter Zumthor in Switzerland. He has designed a glazed envelope that is 'thick' and opaque and which undergoes and inversion of its perception at night, becoming a glowing landmark on the lake in Bregenz.

flows and circulation

A fundamental theme that emerges from the conception of the contemporary museum is the search for a strong spatial continuity and the negation of any element that may be limiting or an obstacle. The architectural form must confirm the fluidity of the interior space and its continuity with the exterior, exalting the role of circulation and internal flows, favouring movement and the dynamic vision of the architectural form.
The contemporary museum experiments with a great variety in the articulation of spaces: the museum becomes a sequence of spatial sequences.
In the Centre Pompidou, the large tube of escalators and circulation is located on the exterior of the building and allows for both the exhibition of the mechanism of vertical circulation as well as the aerial view of the city and thus the osmosis and the almost visual co-penetration of the urban space of the historical city.
The selection of projects presented in the special issue of WA World Architecture about Contemporary European Museum is articulated in three interpretative categories, that also feature less recent projects, such as the Kuntshal by Rem Koolhaas, Kiasma by Steven Holl and the Museum by Norman Foster in Nîmes, all of which have marked a new course in museum design research:
The Museum as an Urban Icon (widely treated in the first part of the essay); the second section, entitled Museum and History highlights the relationships between new museum projects and delicate historical environments and more in general the relationship between architecture and history represented by the city itself. In the words of the philosopher Franco Rella "The large city, the metropolis, is already the supreme site of collection, the museum." A category that is tied to the first, but privileged by the comparison between the ancient and the new in generating new architectural form.
The Museum and Landscape collects a series of important projects that investigate, through the museum, a new idea of artificial landscape, confronting broader contexts and landscapes and thus the dimension of open space. They are projects that dialogue with the landscape, with light and chromatic shifts, with the tactile perception of materials and with the void.
The selection contains only one unbuilt project, by the French architect Rudy Ricciotti, the winning entry in the competition for the Mediterranean Museum in Marseille. This project features the interweaving of the theme of the museum with that of the Mediterranean landscape as the physical space of construction. The MUCEM interprets the theme of the Mediterranean as the physical and perceptive creation of a new visual and sensorial landscape located in the port of Marseille. The envelope is a delicate vibratile and transparent membrane that encloses the space of the museum and opens the view towards the city and the sea, erasing the limits between the internal and external spaces of architecture.
In conclusion, the research into museum design appears to be in harmony with the large interventions of renovation of the most important museum institutions in Europe and the proliferation of specialised museums, the so-called thematic museums.The conception of the contemporary museum appears to find some synergy between the two large scales of intervention. On the one hand, the large building, encyclopaedic in its nature, such as the Louvre that continually renovates and enlarges itself (the Islamic art wing will be designed by Rudy Ricciotti, and the addition to the large hall entrusted once again to Pei). On the other, the specialised museums related to a specific argument, where the architectural form tends to profoundly identify itself with the theme. A trend that mirrors, in a more general way, the centrality of public institutions and, in a parallel manner, the growing role of large, private foundations.
Ben Van Berkel and Caroline Bos of UN Studio see the future of the museum in the specialised museum: "The future therefore belongs to the specialist collections that communicate strongly what they are about and in this way stimulate culture far more than generalist collections without direction. It is our ambition to make an environment that simulates contemplation, but by other means than reinforcing a restricted optical field. We want to achieve an intense visual experience by relating the work to the space it is, to other works in its vicinity, and to the outside world. (...) It's not the iconic face that makes a museum great."
The result is that within the current panorama it is possible to think of a more all-inclusive concept of the museum, to be understood not only as the building for the display of 'objects', but as the space of events, or as the 'story' of a specific theme that becomes the cue and stimulus for reflection and further investigation for the visitor.
Something new is represented by the growing interest for the thematic museum, with artistic and scientific specialisations, whose strutture is clearly different from the more general layout of the Louvre, the Prado, the British Museum, etc. They are museums that develop an interest in a specific argument, often 'interpreting' and representing the theme through the architectural form, with the objective of re-discovering the culture and the identity of the site. It is also possible that the project for a specific theme materialises and renders a sentiment physical, a sensation if we consider the museum as a void, a site of reflection, so evident in the intense memorial to the Holocaust designed by Eisenmann in Berlin.

作为城市空间象征的博物馆

Susanna Ferrini

在当代建筑研究领域，博物馆在城市和公共空间配置方面起着核心的作用。

根据真实数据显示，全世界每个星期都有两座博物馆落成并开幕，因此我们可以确信，专门用于文化交流的场所越来越多地成为了这种意识向更广泛人群延伸的标志。

近年来，正是新博物馆的建造突出展示了建筑在转变一座城市过程中的核心作用，建立了一种几乎可以称为城市规划中的建筑隐私的理念。这在很大程度上影响了建筑研究：毫无意外地，过去十年中两座最为著名的官方文化建筑，位于毕尔巴鄂由弗兰克·盖里设计的古根海姆博物馆和位于伦敦由赫尔佐格和德·梅龙设计的泰特现代美术馆，即使是在众说纷纭的情况下也体现了博物馆建筑的核心作用——作为建筑研究中的推动主题的作用。通过这两座建筑所获得的共鸣超越了其所有的专门研究，使我们相信对于媒体和广大群众而言博物馆空间正在演变成为文化碰撞和交流的专有空间。

因此，博物馆越来越多地作为公共交流空间出现，用于表达新的文化层次。博物馆更倾向于以象征的形式“代表”这种公共仪式的活动，而不是将其局限在简单地展示艺术作品上。

同时，在游客和使用者使用当代博物馆的模式方面也出现了新的演变，使博物馆分成了不同的使用层：第一层以快速基本了解为主，旨在仅仅获得有关展示主题的几项关键概念，再进一层之后便要求深刻的认知，在不知不觉中提高了对艺术活动的兴趣，在此对大众而言自由时间以及文化游览的核心作用变得更加突出。

在更多的时候公众/游客变成了其建筑形式的积极主体。通过空间连接形式和材料的选择刺激了他/她的视觉和感观认知。事实上，上文所提到的两座建筑同大部分当代博物馆一样都明确具有以大型内部空间作为入口的特点：在泰德现代美术馆，位于河堤旁边发电厂侧的大型涡旋式大厅所提供的广阔空间大约占据了该建筑物整体体积三分之一，它是全世界最大的室内公共场所；而在古根海姆博物馆，其表面积为3 00平方米，高度为5 0米的中心中庭没有任何作为直接展示区的功能，而是被当作了该城市一个新的公共活动区域，其目的是为了实现该建筑内部和外部人流和通道的绝对连续性。

事实上，一项在现代和当代辩论中被广泛讨论的主题就是寻求博物馆建筑物和其内容物之间的平衡，以及被作为交流表面积集合体的包围建筑的建筑形式和所展示内容之间的平衡。

博物馆已经成为了城市空间的大型象征标志，这种“容器”的流行形象被当作了人流、活动和信息的聚集器，进而被当作了城市内一种可识别的标志。博物馆包容式建筑体被用来提供一种逐渐增强的强度、情感和认知，同新层次的游客和使用者相互作用及其内容之间的简单相互交流所达到的效果结合到了一起。

在很多情况下，博物馆建筑都通过其所提供的力量成功促进了城市空间的发展并决定了其转变方向。此外，因为在城市规划中越来越缺乏信心以及将此规划作为城市现象的起决定作用的因素的看法导致建筑物具有了高度的象征性意义，它还具有很强的城市特征。

毫无疑问，从20世纪80年代开始我们便开始见证了城市介入逻辑的转变；新的大型博物馆项目的正是要影响和指导城市的转变。这种趋势标志着博物馆的封闭、自指式“盒子”，一种与其外界环境毫无关系只是被用于保存艺术宝藏的“背景幕”的形象的结束。这种类型的主要项目包括雷姆·库哈斯所设计的康索现代艺术中心、斯蒂文·霍尔所设计的位于赫尔辛基的基亚马现代艺术博物馆和由让·努维尔设计的位于巴黎的阿拉伯世界博物馆，它们通过建筑的形式本身确切反映了几何形状和城市结构的组合。

今天我们反问我们自己，是什么原因使博物馆的类型在建筑研究中起到了这种核心作用，在更为普遍意义上体现了当代文化的演变？

事实上，博物馆在全世界、在欧洲的扩增似乎没有结束的时候。因而Harald Szeemaan说道，“对今天的建筑师而言修建博物馆就像过去的建筑师修建教堂一样”，导致形成了“在瑞士每9000名居民就有一座博物馆”这种看似荒谬的状况。

另一方面，其正面作用在于艺术不再被封闭、在自指式的自觉高人一等的状态，而且建筑致力于重新建立一种由广大大众分享其目标的风纪，并同游客达到更高层次的“神会”。而反过来，其危险在于这种趋势导致博物馆建筑达到了过度引人注目的地步，在所展示内容方面给建筑形式赋予了太多的意义；其期望博物馆带来“惊喜”的愿望超过了博物馆主题的“感觉”。

在许多中情况下博物馆的形式都被诠释为一种主题，使其成为了一种需要展示的艺术；毫无意外地，由Daniel Libeskind设计的位于柏林的犹太人博物馆即使在展览开始之前便已经有35万多人参观，其内部展览似乎是被“层加”到了对其建筑形式及其所期望传达的感觉之上。

这种将包容式建筑作为博物馆建筑容体的概念的一种可能后果就是：在外部正式界定了一种大“盒子”，但其内部属于“中性”，而博物馆展品在第二期设计，就像是其内部的第二层“皮肤”被冷淡的交叠在一起。

与此显著不同的就是“现代”运动大师们所发展的新型博物馆空间的概念。在Frank Lloyd Wright、Louis Kahn（尤其是在欧洲）、卡洛·斯卡帕和Franco Albini等建筑师们的作品中，展品的概念被同建筑的概念融合到了一起，如维罗纳的Castelvecchio 博物馆和热那亚的Museo del Tesoro Sa n Lorenzo等就具有一些这种意义。在这些项目中，建筑主题和展览的主题结成了不可分割的“整体”，尤其是所涉及到的核心主题就是使用者/游客与所展示内容之间的关系，在这种过程中建筑师成为了一种媒介，一种理解的向导，而永远不会成为一种隔膜或障碍。

博物馆作为强烈形式和视觉强度的标志和标识的作用突出了将博物馆作为一种“中性”空间和将博物馆作为一种“艺术品”这种矛盾体的存在，正如Frank Gehry在其1998年发表的论文中所述观点。博物馆的“中性”主题与Gehry的作品的关系似乎是一种无法实现的谜题：“艺术家们更愿意在一座不同凡响的建筑内展示其作品，它们不想在中性的地方展示其作品”。即使所用概念方法有所相同，Jacques Herzog同样重视建筑物能够确认其本身是一座建筑而不是虚无之物的重要性：“我认为目前已经有足够的具有设计简单的陈列室的博物馆能够满足人们对平凡的需求”。泰德现代美术院以前是一座河堤侧发电厂，它肯定会具有从基本上而言很平和的陈列室，但同时也会具有戏剧性的质量。我们希望能够扩

展大空间和小空间之间的感觉。“

在这些方面，即使在当代设计中人们也希望能够在一定程度上延续创建了“现代”博物馆概念的基础的大师们的研究。其目标可能就是重新恢复建筑和展览之间的协同作用，以使博物馆成为一种能够在内容方面转变和演化的灵活的建筑结构。

当代博物馆的形象因而就成为了一种沉思机制，在技术上具有复杂性和先进性，并在空间连接和功能方面具有充足的灵活空间，能够更新其内容以及多媒体交流系统，以相互结合的形式跟随并确认历史和科学研究日益增长的节奏。

这些新的诠释类别形成了与欧洲当代博物馆相关的建筑研究的特性并可以同新的诠释概念相结合。

将博物馆作为城市的一部分

如果我们必须要找到博物馆的城市概念的原型，毫无疑问它就是位于柏林Sprea河一座岛屿上由Karl Friedrich Schinkel于1824~1830年间设计的Altes博物馆。Altes博物馆是德国第一个公共博物馆中心和雏形，也是第一座从刚开始便作为一座博物馆来建造的公共建筑。建造博物馆群岛的努力持续了近乎一个世纪，根据一项计划建造了一系列的相关建筑，包括Neues博物馆、Alte Nationalgalerie、Bode博物馆和Pergamon博物馆，用于展示在希腊Pergamon 发现的大型Altar到Zeus。

Museumsinsel，现在已经被完全翻新，是一个复杂的博物馆城系统，它能够坚定地赋予城市环境独特的特点，这一性质也可以在维也纳由Ortner&Ortner设计的Museumquartet中看到。

因此，博物馆城本身属于一种城市基础设施，是城市传播和流通的场地。

另一方面，在欧洲，通过设计大型历史城市突出了文化复兴运动的发展，以外科手术形式插入到历史城市结构中的“新建筑”，首先就是通过修建从一开始便被当作城市本身的标志和标识的新博物馆开始的。

新的欧洲博物馆中心正是用于满足日益变得更具选择性和多样性的文化旅游的新要求的转变和变化的主体。P ei在巴黎的L ouv re所使用的玻璃金字塔以及由Rudy Ricciotti设计的新增伊斯兰艺术翼台；Uffizi所面对的由Arata Isozaki设计的大型入口Arata Isozaki重要主题；泰德现代美术馆已经在计划一项由赫尔佐格和德·梅龙设计的扩建项目。

最近，博物馆项目所采取的策略正是指导城市转变的策略。

很多最为重要的欧洲城市，其中包括巴黎、伦敦、柏林、维也纳、毕尔巴鄂和罗马，在最近几十年中都决定通过大型博物馆项目及扩建项目来重新塑造其形象。博物馆似乎证明了一个国家的政治和文化生活，是在海外宣传其特性的理想主题。

20世纪90年代末的标志性案例是毕尔巴鄂的古根海姆，此项目成功衬托出了其所在整个区域的经济和文化发展，我们甚至可以将“毕尔巴鄂效应”用作发展的代名词。

当今人们对新博物馆的期望是什么呢?

实际上通过某一项单一的构思或概念来描述博物馆几乎是不可能的。

博物馆研究中心所公布的当代艺术研究似乎越来越多地和整体文化现象，以及大众和艺术文化场地研究方法上的根本变化结合到了一起。使博物馆具有了一种归属于整体社区的感觉，成为了共享交流网络的一部分。

但是，博物馆也是一个消费场所，其消费的是文化形象，是获取共享交流网络密码的万能钥匙。这种意识通常非常基础且紧迫的，在这里我们希望以整体社区代表的形式参与进来。这是在选择建筑形式时必须要面对的一种愿望，而且必须对此问题作出答复。

多问机制

如果我们愿意，我们可以将博物馆当作是对多种复杂多样的空间和功能的组织，来以数字比喻的方式了解多问机制的体验和效果。博物馆成为了一种用来诠释信息社会某一特性的机制，这种特性可以通过以平均的关注程度在压缩的时间内与不同的经历协同共存的方式发现。因此，多问机制所面对的就是一种与当代社会紧密联系在一起的新型体验。

在更为普遍的意义上，博物馆毫无疑问地已经转变成为了一种复杂的系统。博物馆和档案馆，一种进行研究和科学发布的场所，因而在其所计划的经历层次需要采用多问机制。

其起源毋庸置疑就是Centre George Pompidou，一种使用于举办“活动”的暂时性展示空间、容纳展品的房间、大型图书馆和档案室、多媒体空间、会议和投影室以及包括书店、饭店和咖啡厅在内的商店和休息区得以“共存”的机制。

全球博物馆网络

大众文化旅游的新活力和传播促进并生成了新的博物馆项目，不仅仅是在国际大都市中，在较小的艺术中心也同样如此，最主要的就是在那些正在经历转变以寻求创建其新特色的城市区域。

当描绘受媒体和信息网络所支配的当代社会的新前景时，Paul Virilio提出了从大众社会这种在20世纪后半叶盛行的现象到大众个体化的转变，后者的特点就是通过网络以类似毛细血管的方式传播一种全球化的行为，即使是同个性感知和经历结合在一起时也是如此。

此研究令人神往的一个方面就是将博物馆当作是一种公共区域，同时也是具体游客一种个性经历场所，一种获取能够同所展示艺术作品产生关联的个性空间的场所。

欧洲博物馆的理念必须要面对大众文化的影响：作为这一趋势的一部分，由博物馆所提出的要求被进一步细分，以增加文化所能提供的内涵并进一步限定这些内涵的性质。

由此，将博物馆网络的扩增确定为未来发展前景组成部分，而古根海姆博物馆则被当作了其象征性范例。在世界范围内建造了各种各样的博物馆：包括在纽约、威尼斯、拉斯维加斯、萨尔茨保、柏林、毕尔巴鄂等等。Thomas Kerns认为：“古根海姆的理念就是国际博物馆的理念，以其不连续的展示空间为特色，传播到了多个国家。我们的批评家们称我们各种各样的博物馆为‘特许经营店’，或将它们叫做‘麦当劳’连锁店，但这种说法完全不得要领。这种‘特许经营’的模式暗示着需要完全的标准化，彻底不管当地的背景；而古根海姆在扩建其博物馆网络的时候则采取了正好相反的方法。没有任何两种情况是完全相像或在总体上一致的”。这些话语突出了介入设计的新策略：博物馆位于各个国家中，其建筑物具有强烈的背景和主题特色。博物馆网络的模式将增加当前社会中典型的文化“漫游”。

作为包容式建筑的博物馆

博物馆在城市中的重要性及其作为重要文化"吸引区"的作用使人们将其认定为了一种特别的包容式建筑。

通常在设计这种包容式建筑的时候，人们希望能够突出强调其作为城市背景中建筑物的作用，使其具有很高的物质性，这就与其所能够容纳的展品，即其给人们带来惊喜和惊诧的能力之间出现了矛盾之处。

通过采用新的形式和使用新的材料，以及寻求更"大"的规模等方式，突出显示了将博物馆建筑作为"精心设计的展览"的这种趋势。而在许多项目的形成过程中已经将博物馆作为了一种能够在规模和完整性，以及它能够为文化、研究和旅游带来的多样化模式方面自给自足的"大本营"。

尽管如此，集中建造包容式建筑这一主题的努力还是被当作了一种为其外部"表层"寻找新物质性的有趣研究。通常这种包容式结构都被解释成了博物馆和城市空间之间的"界面"，成为了其内容物进行沟通交流的系统。在赫尔佐格和德·梅龙为XX世纪艺术博物馆所设计的项目中，外表面被点缀上了特色性的数字图像，描述了该博物馆未来在内容方面随时间的演化。

在由彼得·卒姆托设计的位于瑞士的博物馆中即可发现这种对高度物质化和意义的追求。他设计了一种"厚重"且不透明的光滑包容式建筑，由此体现了他对夜晚的理解，它成为了Bregenz湖畔一座耀眼的标志性建筑。

流动和循环

当代博物馆中所衍生出的一种基本主题就是对不同凡响的空间连续性的追求，以及对任何可以成为限制或障碍的成分的摒弃。建筑形式必须能够巩固内部空间的流动性及其与外部环境的连续性，提高循环和内部流动的重要性，偏重于建筑形式的移动性和动态视角。

当代博物馆在空间连接方面进行了多种多样的实验：博物馆变成了空间序列之一。

在Centre Pompidou，其大型的升降梯和循环管路位于建筑物的外部，既可以向人们展示其竖直循环的机制，又可以使人们从高空中欣赏到城市的美景，从而使其渗透到这座历史城市的城市空间之中，甚至与其相互交融为一体。

在此问题方面所提供的项目示例采用三种诠释类别进行连接，这些项目也都不是近期的项目，如雷姆·库哈斯所设计的康索现代艺术中心、斯蒂文·霍尔所设计的基亚马现代艺术博物馆、Norman Foster所设计的位于N·mes的博物馆，它们都标志着博物馆设计研究中的一种新阶段：

作为城市标志的博物馆（在本文第一部分进行了广泛的论述）；第二部分题目为博物馆和历史，着重论述了新博物馆项目和历史环境之间建立的关系，在更为普通意义上即是建筑和城市本身所代表的历史之间的关系。如哲学家朗哥·雷拉所说，"大城市，国际大都市，已经发展成为了收藏品的重要集聚地，即博物馆"。这种类别同第一种结合在了一起，但是通过在生成新建筑形式方面将过去和现在进行了对比使其具有的一种独特特色。

博物馆和景观收集了一系列通过博物馆研究新型人造景观的重要项目，这些景观所面对的是广泛的背景和风景景致，也是开放型空间的维度。它们是可以通过光和色彩的漂移，通过对物质材料的触觉感知，通过空间和风景景观进行对话的项目。

在所选项目中只有一项未建项目，此项目由法国建筑师鲁迪·里齐奥蒂设计，它是马塞地中海博物馆入口设计竞赛的获奖作品。此项目突出了博物馆和作为建筑项目物理空间的地中海式地貌之间相互交融的主题特色。此MUCEM将地中海主题诠释为一种物理和感知创造，在马赛港港口营造了一个崭新的视觉和感觉景观。其包容式建筑是一种精致具有颤动特征的透明膜，容纳了整个博物馆空间并在城市和海洋景观方面保持开放，消除了该建筑内部空间和外部空间之间的界限。

总之，博物馆设计的研究似乎同欧洲大部分博物馆研究所的大型翻新项目以及专业化博物馆——即所谓的迅速增多的主题博物馆的情况——协调一致。

当代博物馆的概念似乎在两个大规模介入项目之间找到了一些协同效果。另一方面，如卢浮宫等大型、百科全书般的建筑持续进行翻新扩建（伊斯兰艺术侧厅将由鲁迪·里齐奥蒂设计，而其大厅的扩建项目则将由贝律铭来完成）。另一方面，与某一具体论题相关的专业化博物馆的建筑形式在很大程度上会与其主题相匹配。这种趋势在广泛意义上反映了公共研究机构的核心作用，同时也突出了日益增长的大型、私有基金会的作用。

UN工作室的本·范·伯克尔和卡罗琳·博斯认为博物馆的未来在于专业化博物馆，"因而未来属于那些专业化的收藏，它们能够同其所属领域进行有力的交流，这样比没有方向的大众化收藏在促进文化发展方面的作用要大得多。我们的抱负在于创造一种能够促进研究的环境，但不是通过能够增强有限视野的方式。我们希望通过将作品同其空间、同其附近的其他作品以及通外部世界之间建立起联系的方式提供一种强烈的视觉体验……一座博物馆之所以伟大并不是因为其标志性外观。"

其结果就是，在当今的整体环境之下，我们可以考虑使博物馆具有更加包罗万象的特性，使其不再仅仅是一座展示"物品"的建筑，而是一个举办活动的空间，或是一个具体主题的"故事"，而它又能为游客提供反思和进一步研究的线索和促进因素。

根据新的信息，人们对在艺术和科学方面具有专业化特性的主题博物馆的兴趣越来越大，其结构明显与如卢浮宫、普拉多和大英博物馆等整体布置模式不同。

它们是兴趣在于某一具体论题的博物馆，通常在其整体建筑形式中都在"诠释"和表述这种主题，其目的在于重新发现其所处场地文化的特色。具有具体主题的项目还可以物质化，使某种情感具有具体的形式，即如果我们将博物馆看作是一个空间、一个反思的场所时所能感觉到的感觉，位于柏林由Eisenmann设计的Holocaust博物馆就给人一种强烈的纪念感。

Capturing the blue from the sky

Museum Design between Art and Architecture
The Relationship between Container and Content in Museum Design

by Antonello Stella
from WA-World Architecture n.195 2006, Tsinghua University, Beijing

"I want to capture the blue from the sky" are the words used by Carlo Scarpa (1906-1978), one of the masters of Italian architecture and designer of the some of the most interesting museums in Italy, to describe his project for the addition to the Gipsoteca [Plaster Cast Museum – TN] dedicated to the sculptural work of Canova and located in Possagno in Veneto in north eastern Italy.

His statement, simultaneously synthetic and dense with references and meaning, contains one of the principal themes of museum design: conserving and protecting the objects on display from direct sunlight and exterior conditions without, however, entirely negating the relationship between the museum's contents and the external world, seen not only as space but also as thought. Thus the space of the museum is not a "human scaled" space but rather, operating by metaphor, a medium between mankind and history where architecture assumes an evocative meaning. The architectural container thus assumes particular meanings: light itself, from the exterior and properly filtered, must enter the interior space of the museum and, at the same time, the object on display must make reference to a broader cultural context that extends beyond the museum itself.

It is possible, without a doubt, to state that the space of the museum is a continuous flow between interior and exterior space, where the building-container, and thus the architecture of the museum, becomes the expression of this dichotomy. However, more in general, by broadening the meaning of Scarpa's words from the realm of space to the more theoretical realm of the meaning of architecture, it is possible to state that the museum becomes the space where the dichotomy between art and architecture, so highly discussed by architectural critics during the last century, is most clearly highlighted. It is precisely within the museum that architecture demonstrates that it is not simply something "technical".

In an architectural magazine created for an Eastern audience this reflection acquires further meaning: in fact, it is well known that one of the principal differences between Western and Eastern culture is precisely our means of perceiving objects, events and, more in general, the entire would via couplings of opposites, while the Oriental vision of the world privileges the unity of vision; in short, we Westerners think of mankind and the world as separate entities, while for those of the East man is not separate from the world. We can say that it is not man who observes his history, as if they were two separate entities, but man in his history. The comparison between these two cultures through architecture may lead to interesting reflections, assisted by the fact that architecture shifts the level of confrontation from the technical and technological realm, perhaps less interesting, to the cultural realm. This reasoning about the museum renders the technical question, as necessary as it is to architecture, somewhat secondary.

As a result, if for us Westerners the question of the dichotomy between art and architecture has fed the discussion related to this argument during the past century, this is not in the case in the East, where this discussion did not have, nor does it have, any reason to exist.

The concept of the Museum raises this dichotomy. Is the museum not perhaps a space that contains art, and presents itself as an artistic-architectural object? We will see later how the "artistic content" of the museum assumes continually diverse connotations.

Many of the examples published here are not "anonymous boxes", designed essentially as functional containers for works of art: architecture at the exclusive service of art, where the display space is designed exclusively based on light, colours, the flows of visitors who, from the exterior, filter into the interior space and the optimal conditions for the display of art.

These are not the architect's only concerns. Though they are fundamental elements for the design of the project, they are not the only ones. Museum architecture aspires to become, here more than in other architectural typologies, "a work of art".
If the museum is the building that most highlights the problem of the relationship between art and architecture, the examples presented in this issue, which represent the most important works of contemporary architecture, appear to indicate that the distance between the work of art and the work of architecture continues to decrease.
In the majority of the projects the desire of the work of architecture to show itself as a "work of art" is very strong, as demonstrated by the fact that the envelope does not wish to be as anonymous as possible; if the architectural lesson of the Modern Movement, of which Scarpa, even with his own highly personal language, is a continuation, tended to interpret this relationship by reducing architectural expression to a minimum (for Scarpa only light), for contemporary architects, the architectural envelope, even while respecting and searching for a correct relationship with the artistic content and the "necessities" of displaying them within the space of the museum need not renounce its own autonomous level of expression. This is not simply a formal intention: the will to artistic expression must not be confused with a sterile will to form. Contemporary architecture has absorbed the most recent trends of artistic expression - less "representation" and more "performance"; continually less form and ever more event. This is why many of the examples published here, even, and perhaps above all, those that apparently aspire to a reduction to the minimum of the formal expression, show a strong desire towards expression through the matter of architecture. The museum building itself becomes the object to be perceived and is made manifest by showing, in an anything but anonymous manner, the matter of which it is constructed. It displays itself as a work of art even before displaying its contents. It aspires to be a well known and recognisable sign within the landscape, whether it be natural or urban.
Scarpa's building in Possagno is a masterly example of Museum architecture as an interpretation of the tension between exterior and interior in both physical and symbolical terms: physical exterior space, represented for Scarpa by sunlight, enters into the building not through simple windows, but through angular cuts in the volume (in both the vertical and horizontal surfaces) that, as stated by Scarpa himself, wish to merge the surfaces of the building with the ideal plane represented by the sky itself. The backdrop of the interior walls is an anonymous as possible and is called upon to offer the maximum effect for Canova's sculptures. In reality, Scarpa's intention is precisely that of cancelling the building, as if the works existed in the open air, even while protected against the elements and surrounded by solid walls. However, if we analyse many of the "envelopes" shown here, they appear to interpret the tension between interior and exterior, container and content not as the cancelling or reduction to the minimum of this relationship, but as a relationship made of continuous connections. Without the envelope, contrary to the museum in Possagno, the display space in these museums would probably appear to be lacking a fundamental and essential part of the perception of the whole.
If we return now to the parallelism between the Western and the Eastern approach, as part of the attempt to generate a reflection aimed at creating a dialogue between the European examples shown here and which represent an experience that is the result of a lengthy process begun in the 1800's with the display pavilions that contained lengthy sequences of sculptures with the possible future trends of a museum in China, it is possible to imagine a future contribution to museum architecture in China through the approach described in this essay: Chinese architectural research, in particular the relationship between container and content and art and architecture in contemporary terms, in one context, above all the urban, is very active.
The great deal of architectural activity that is currently leading China to a significant level of involvement in construction and reconstruction of considerable portions of the residential and tertiary fabric within cities now finds itself carrying out a necessary reflection about the conservation of its enormous historical patrimony, from the most ancient to the most recent.
This reflection acquires further meaning in relationship to

the experience of two years of cultural exchange between China and two European countries: France (2005) and Italy (2006). These years have been witness to the intensification of exchanges between two cultures, the European and the Chinese, above all in the realm of history-art.
The recent display of a selection of principal works of Italian art from the medieval period to the 19th-century, with particular focus on the golden era of our civilisation, the Renaissance, at the WAM in Beijing, where we were involved as the designers of the exhibition, presents itself as a starting point for a shared reflection about the sense of the relationship between the work of art and architecture, between the container and content, as part of an approach that, however, takes into consideration the exchange and the intersection between two historically different cultures as part of a perspective that is increasingly focused on exchange. The work carried out in collaboration with the management and those responsible for the WAM highlighted just how vast the possibilities of collaboration and the exchange of views in this field truly are.
This experience is continuing, again in relationship to the year dedicated to the Italy-China cultural exchange, with a second exhibition dedicated to the "Great Civilisations", whose broader approach represents an opportunity for ever more stringent comparison. The show will be held in October 2006
The idea of separation and juxtaposition between opposite poles, typical of Western thought, perhaps requires a reconsideration that could possibly begin with the more unitary nature of the Eastern approach; in the same way that the experience developed during the reflection on the relationship between art and architecture and, more in general, between man, the work of art and its historicisation may be useful for the future of the Chinese experience in the realm of museum design.
What we propose here is to look not only to look at these projects as works of architecture and the result of experience matured in Europe in the field of museum architecture, which we hope is well represented in the examples published here, but rather to observe these works of architecture as part of the approach being described here, which goes beyond the simple fact of architecture and the images shown.
The necessity of reasoning beyond the pure data of architecture is also supported by a perspective that is only partially described in this issue, through only a few examples, but which will become, in the future, ever more central in the field of museum design and construction: the content of museums in the next century will be increasingly more "immaterial", as a consequence of the evolution of mankind that is witness to the increasing dominance of the immaterial over the material. For the spaces of the museum this means that the history of the object exposed will become of ever greater importance. The content will be increasingly more representative, described using the most sophisticated digital technologies, and the container – the architecture – ever more evocative.
The idea of uprooting the work of art from its context in order to display it in another, enclosed space is definitely coming to an end. It is the Western idea, the romantic and nineteenth century idea of uprooting the work of art to construct, elsewhere, a monument for its conservation and valorisation. In China the experience of the large Museum dedicated to the clay soldiers of Xian demonstrates how this data has been acquired and how it is important even for the socio-economic development of the context within which a cultural patrimony is located.
On the other hand, beginning with the assumption mentioned at the beginning of this essay – that the museum is the medium between mankind and his culture – the paradigm of the museum as a container of objects is about to exhaust itself, creating space for the museum-story. The museum of the future will be more a space for events, where the production of spatial and perceptive sensations is more important than the physical presence of objects. The virtual will continue to invade the spaces of the museum.
This passage may allow for the overcoming of the content-container dichotomy and, more in general, that between art-architecture. The museums presented here, which focus on the most advanced research in the realm of museum architecture and methods of displaying works of art, already represent a point of passage between the old and the new paradigm: they are a bridge and thus a mix of the two concepts, not yet entirely immaterial and no longer solely a container of objects. As the story becomes ever more important with respect to the object, the two cultures, the Western and the Eastern will perhaps be able to find more common points of overlap. Often the appearance of the forms differs more than the stories of those who have generated them. In this sense, as mentioned, the forms - the result - become less important with respect to

the will to expression and thus the process that generates them. The historical process becomes more important than the result.
This attitude of taking a distance from forms as a result protects that which is in inherent in this reasoning and which aims to bring the two cultures closer together against the possible risks of "globalisation", without obviously losing the unique qualities within a generalised simplification.
It is precisely in the specific realm of museum architecture, more so than in other typologies, that the comparison between these two cultures may generate new horizons of research and open new perspectives for all involved.
The museum presents itself as the site of difference, of identity tied to culture and place, but which, thanks to the great immaterial network that connects all of us, allows for a comparison of these differences as part of a logic of exchange of ideas at the global level. The space of the museum expands out of scale and becomes, in fact, a living organism.
Scarpa's lesson, to which this essay belongs, is located at the extreme opposite of the conclusions to which we have come. His museum designs, which made each object on display an autonomous story and which interpreted the museum building as the direct emanation of the story, lead us to believe that the captured piece of the sky that he wished to create may be the starting point for the reconfiguration of the two opposite poles of this thesis.
This is also true because the blue of the sky in Beijing or Shanghai is not so different from that in Rome or Paris, even given the diversities of those who observe it. We must also remember that ever more residents of Beijing are looking at the sky in Rome, and a growing number of Romans are looking at the sky in Beijing.

捕获天空中的蔚蓝

在艺术与建筑之间的博物馆设计
博物馆设计中容器和内涵的关系

Antonello Stella

意大利最为有趣的博物馆设计者和建造者建筑大师卡洛·斯卡帕(1906-1978)曾经说过"我想捕获天空中的蔚蓝"，他就是用这句话来描述他所执行的Gipsoteca [Plaster Cast Museum-TN]扩建项目的，此项目位于意大利东北部Veneto的Possagno，专门用于展示卡诺瓦的雕塑作品。

他的话语在参考和意义方面同样综合、致密，它包含了博物馆设计的一个主要主题：保存和保护所展示物品不受阳光直射和外界条件影响，但同时又不能够完全否定博物馆内展品同外部世界的关系，不仅是空间同样也是思想之间的关系。因此，博物馆空间不是一种"可界定的"空间，按照比喻的说法它是人类和历史之间的中介，其建筑物具有一种唤起回忆的意义。

这样这种建筑容器就具有了特殊的含义：从外部进来并经过适宜过滤的光本身必须进入博物馆的内部空间，同时所展示的展品必须参考超出博物馆本身的广泛文化背景。

毫无疑问，我们可以说博物馆空间是一种内部空间和外部空间之间的持续流通，而其建筑容器，即博物馆建筑，成为了这种矛盾体的表现形式。

但是，在更广泛意义上，通过Scarpa所说的话的意义从空间的领域延伸到建筑更为理论化的领域，我们便可以说博物馆已经变成了艺术和建筑这对在20世纪被批评家们充分讨论过的矛盾体最为突出的场所。建筑物正是在博物馆中证明了它们不仅仅是一项"技术性"事物。

在一份面向东方人的建筑杂志中，这种反映获得一些新的意义：实际上，众所周知西方文化和东方文化间的主要差异正是我们认知事物、活动，更广泛地说认知整个世界的方式是将对立的事物组合在一起，而东方人眼中的世界更注重视觉的统一性；简而言之，西方人认为人和世界是相互独立的实体，而东方人则认为人和世界是密不可分的。我们可以说，不是人类在观察历史，而是人类处于历史之中，就像二者是两个独立的实体一样。

如果通过建筑比较这两种文化我们可以发现一些有趣的反映，而建筑可以影响技术领域和文化领域之间的冲突，虽然这一点可能不如上一点有意义。我们对博物馆的评论使得这一技术问题似乎变得有些次要了，虽然它对建筑物而言同样重要。

结果，如果对于西方人而言艺术和建筑这一对矛盾体在过去的一个世纪中促进了与此论题相关的讨论，而在东方则完全不同，因为这种讨论过去和现在都没有存在的理由。

正是博物馆的概念导致了这一对矛盾体。难道博物馆不仅是一个里面可能收藏有艺术空间，而其本身可能也是一件具有艺术特色的建筑物？我们在下文将讨论博物馆的"艺术内涵物"是如何继续拥有多种多样的内涵的。

毫无疑问，此处所举的示例即可以回答此问题。

许多此处所发表的示例都不是本来即设计作为装盛艺术作品的功能性容器一样的"匿名盒子"：建筑物专门为艺术而服务，展示空间专门根据光线、色彩和游客流量设计，它们可以从外部经过过滤进入内部空间，以达到最佳的艺术展示条件。而建筑师们所关心的并不仅是这些。虽然它们对于项目设计而言是非常重要的因素，但并不是仅有的因素。博物馆建筑比其他任何形式的建筑更渴望能够成为一种"艺术作品"。

如果博物馆能够成为最能突出艺术和建筑之间的关系的问题的建筑物，而此处所提供的能够代表当代建筑中最为重要作品的示例似乎能够表明，艺术作品和建筑作品之间的距离正在持续缩小。在大部分项目中建筑作品希望体现其"艺术作品"特性的愿望非常强烈。这可以通过包容式建筑希望能够尽可能地避免匿名的状态而看出来。对于建筑学的现代运动而言，斯卡帕便是这种运动的继承人，即使用他自己的话来说也是如此，如果这种运动试图通过将建筑的表达形式降到最低的方式（斯卡帕只采用了光线）来解释这种关系，对于当代建筑师们而言，建筑包容体也不需要放弃其自己的表达形式，即使是需要遵守和寻求与其艺术内涵之间的正确关系并"需要"将其在博物馆空间中进行展示。这并不仅仅是一种形式上的意愿：不能将艺术表达的意愿同单纯的形式意愿混淆在一起。当代建筑吸收了最近的艺术表达形式——少一些"表现"，多一些"业绩"；更少的形式和更多的活动。这就是为什么此处所发表的许多示例都强烈希望通过建筑事物来表达自己，甚至是那些明显希望将形式表达减少到最低程度的示例也是如此。博物馆建筑本身成为了需要认知的对象，并通过展示来显示其所构建的事物，但决不是通过匿名的方式。即使在展示其展品之前，它便已经作为一件艺术作品展示在了人们的眼前。它希望能够成为无论是自然还是城市地貌风景中一道众所周知、清晰可辨的标志。

Scarpa在Possagno所建造的建筑就是博物馆建筑在物理和象征两种意义上诠释内部和外部之间压力的一项大师级范例：物理外部空间在斯卡帕的作品中通过阳光来表示，阳光不是通过简单的窗户进入建筑物之中的，而是通过该建筑体的一种具有角度的切口（水平面及竖直面上均有）进入，正如斯卡帕自己所说，他希望将这些建筑物表面同天空所代表的理想平面融为一体。其内墙的背景被设置得尽量没有任何个性特征，以最好地凸显卡诺瓦的雕塑作品的效果。在事实上，斯卡帕的意图正是要取消建筑的形式，即使艺术作品实际是被重重保护而且被包围在坚实的墙壁之中，也就像是艺术作品被直接展示在露天之中一样。但是，如果我们对此处所示的众多“包容式建筑”进行分析可以发现，它们解释内部和外部、容器和内容物之间压力的方式似乎不是通过取消或尽可能降低这种关系，而是将其当作是一种具有持续关联的关系。如果在Possagno博物馆没有同其形成对比的包容式建筑，那么这些博物馆中的展示区域可能就会显得缺少了作为一种整体的关键的基础部分。

现在我们重新返回有关西方模式和东方模式之间的平行论，作为旨在使此处所示的西方示例和中国博物馆将来可能趋势之间进行对话的构思努力的一部分，此处所示西方示例代表了自1800年以来便已经开始的漫长经历，其展示亭中充满了漫长的系列雕塑。通过这种讨论，我们可以预见如何能够利用本文中所述模式来为中国未来的博物馆建筑的发展作出贡献。中国建筑研究在城市背景下显得特别活跃，尤其是在容器和内容物之间的关系以及当代意义上的艺术和建筑之间的关系方面。目前，中国正在进行的大量建筑活动使中国大量参与到了对城市内部很大一部分住宅和三级建筑物的建造和重建工作中，而现在这些建筑活动也开始进行一项必要的考虑，那就是如何来保存其巨大的从最古老的到最现代的历史遗产。

在同两个欧洲国家，法国（2005年）和意大利（2006年）进行了两年的文化交流之后，其所得到的经验使得这种构思获得了进一步的含义。

这两年时间见证了欧洲文化和中国文化之间交流的逐步增强，最为重要的就是在历史／艺术领域。

最近，在北京的WAM展出意大利从中世纪到19世纪主要艺术作品，此次展览尤其集中在我们的文明的黄金时代——文艺复兴时代。而我们作为展会的设计者参与了这项盛事，我们将此次展会当作是一个起点，来分享我们对艺术作品和建筑作品之间、容器和内容物之间的关系的理解。但同时也考虑到了两种历史完全不同的文化之间的交流和交叉，而我们的观点越来越集中在了交流方面。我们同WAM管理层和负责人共同合作所完成的工作恰恰突出显示出了在此领域进行合作和交流能够带来的巨大的可能性到底有多少。

这样的经历还在继续，同样是与意大利／中国文化交流年结合在了一起，第二次展会集中在“伟大文明”，其更广泛的方法使我们能够进行更具有说服力的比较。此次展会于200 6年10月举办。

两个对立极的分离和并列的观念是典型的西方思维模式，我们可能需要从更为整体的东方思维模式开始对其进行重新考虑；同样，在考虑艺术和建筑之间的关系，乃至更广泛意义上的人、艺术作品及其历史之间的关系时，所获得的经验可能对于中国在博物馆设计的未来发展有所益处。

我们在此提倡的是，不仅要将这些项目看作是建筑作品，看作是欧洲在博物馆建筑领域已经成熟的经验的结果，我们希望此处所发表的示例已经很好地说明了这一点，而是将这些建筑作品看作是此处所述方法的一部分，超出了简单的建筑事实及其所展示的形象。

本文中只通过几个示例所阐述的观点也支持了这种不仅要对建筑的纯数据进行论证的需要，但这种观点在将来在博物馆设计和建造领域将起着更加核心的作用：这种观点就是，随着人类进化过程中非物质成分相对于物质成分越来越地占据了统治地位，博物馆的“内容物”在下个世纪将变得越来越“非物质化”。对于博物馆空间而言这就意味着物品所在的历史相对于以前将具有越来越高的重要性。其内容将越来越具有代表性，需要采用最为复杂的数字技术进行描述，而其容器——建筑物——从未像现在这样显得如此重要。

将艺术作品从其背景中拔起放在另一种背景中，一种封闭空间中，展示的想法已经明确到了尽头。将艺术作品拔起在另一处重建，以一种纪念碑的模式进行保存和限定的想法是一种西方的、幻想的、19世纪的想法。在中国，建造专门用于保存西安兵马俑的大型博物馆的经验已经证实了这些数据是如何得到的以及它的重要性，甚至是对某一文化遗产所在的社会经济发展背景的重要性。

另一方面，根据我们在本文开始所提到的假设，即博物馆是人类和其历史之间的中介，博物馆作为物品的容器的范例这种说法即将不攻自破。未来的博物馆将更多地是一个举办活动的空间，其所产生的空间和认知感觉比物品的物理存在更为重要。虚拟成分将继续侵蚀博物馆的空间。

这种转变可能会克服这种内容物–容器矛盾体理论，即更普通意义上的艺术–建筑之间的矛盾体理论。此处所示的博物馆都集中在了在博物馆建筑领域和展示艺术作品方法方面最为先进的研究，它们已经在一定程度上反映出了这种新旧模式的转变：它们是这两种概念的桥梁，因而也就是它们之间的混合体，还没有完全非物质化，但也不再仅仅是物品的容器。

随着有关物品的故事变得越来越重要，这两种文化，即西方文化和东方文化，可能会发现更多的共同点来进行交融。通常，形式的外观之间的差异不仅仅存在于生成它们的故事里面。在这种意义上，如上所述，其形式、其结果相对于希望进行表达的意愿及其生成的过程而言的重要性要差一些。历史过程变得比结果更为重要了。

这种同形式之间保持距离的态度的结果保护了这种论证中的固有成分，旨在拉近这两种文化之间的距离以防止出现“全球化”的可能，但同时在通用简化之后并不会明显丧失其独特特质。

同其他形式的建筑相比，这两种文化的比较在博物馆建筑这一具体领域中更容易生成新的研究领域，为所有涉及到的方面展开新的前景。博物馆本身就是展示不同文化，并和地点结合到一起的场所，但是因为有了将我们所有人联系在一起的巨大非物质网络，使我们能够在全球的范围内交换意见并对这些差异之处进行比较。博物馆空间扩增到了其范围之外，实际上已经成为了一种活体。

斯卡帕的课程，也就是本篇论文所述的课程，与我们所得出的结论正好相反。他的博物馆设计使得每一件所展示物品都成为了一个独立的故事，将博物馆建筑诠释成为了这种故事的直接源泉，这些设计使得我们相信，他所希望捕获的那片天空可能就是对本文两个相对极重新进行配置的起点。

这一点同样适用，因为北京或上海天空中的那点蓝色与罗马或巴黎的没有什么区别，即使进行观察的人各不相同也是如此。我们还必须记住，越来越多的北京居民开始观察罗马的天空，而更多的罗马人也在观察着北京的天空。

Inteview to n!studio

Museum Design between Art and Architecture

The Relationship between Container and Content in Museum Design

by Antonello Stella

from Italy now? , Cornell AAP Publications, New York 2007

how do you define architecture? What does it mean to practice architecture today?

Augustine would say that defining architecture is like defining time: if no one asks me I know it, but if someone asks me I do not know it. It's right in front of us, and every description can seem obvious. However, if we try to make some sense of its underlying meaning, every descriptive attempt always seems insufficient. The good work of architecture is, perhaps, what immediately after its purely "visible" aspect manages to show its "sensitive" aspect, and this is apart from its simple appearance.

One could say that architecture today is still laying inside the paradigm "of the apparent form" that in other fields such as philosophy or art, for example, is being overtaken or is certainly in crisis. A generalized sense of inadequacy of actual architectonic form is evident when comparing new modalities for living in the cities and in the territory to the acceleration of the radical transformation that is taking place. In this sense we don't believe in a unique formal or linguistic reply, elaborating more articulated and spectacular architectonic forms, when it's clearly necessary to develop concepts of flexibility, transformation, growth, temporality of architecture, rethinking the same tools of the project to the point of asking oneself about the same statute of its very existence and legitimacy. The architectonic project needs to necessarily bear in mind the new centrality of the subject, spreading itself over mass individuality, whereby the vision of the individual becomes fundamental as well as the possibility of a true process of appropriation and transformation of architectonic space. Therefore, and we now come to the second part of the question, practicing architecture today has more to the new centrality of the subject, spreading itself over mass individuality, whereby the vision of the individual becomes fundamental as well as the possibility of a true process of appropriation and transformation of architectonic space. Therefore, and we now come to the second part of the question, practicing architecture today has more to do with ethics than with aesthetics. Architecture today is, unfortunately, inside an aesthetic paradigm, opposite to a philosophical and artistic trend that researches new "ethical" horizons. Architecture needs to research not only "the shape of things" but above all a genetic code that will guide and encourage "the relationship between things themselves," as its transformation and evolution, up to even forecasting its possible "disappearance."

does Italian architecture exist? How can it best be spoken about? What best characterizes it?

It's not possible to talk about Italian architecture today as it is to talk about Spanish, Dutch, French, or American architecture. The substantial renouncing action of the architectonic culture during the 80's–90's, with rare exceptions that worked, above all, outside our country, pursuing a continuity of the modern and featuring technical and technological evolution, generated a loss of weight compared to the foreign enviroments. Wheras in other countries identity has been build on a solid body of architectures that was speaking, for better or worse, to their evident "reality", Italy, in its almost obsessive concern to find its own "linguistic identity", built a real self-referenced "cage" around itself. A certain non-project culture, concealed by a presumed "minimalism" of nonintervention, is recently spreading or even re-asserting itself around the country; this would mean to dangerously reset to zero all the attempts made in the last ten years by more than a few architects of our generation. In the last decades, the separation between the "professional way" and "cultural way" of the architectural project and, therefore, between profession activity and university carrier, has been furtherly influencing italian architectural landscape. On the contrary, an aspect that seems to characterize part of the youngest Italian research and in which we recognize ourselves, is the willingness to face head-on the real world of construction, faced with the same density of theoretic research that has always distinguished architectonic thought in Italy. This is certainly outside of an istitutional aknowledgement of schools, far from an obsessive wish of expressive linguistics. As already expressed in the first answer, this attitude also has to do with a new "planning ethic."

new, stable, comfortable, shared, recognizable, exciting, beautiful . . . What values and objectives do you refer to when designing your architecture today, and why?

We believe that the thought, intended as "sense" of the

project, is more long-lived than its form and only through it there can be a common and collective experience of research such ours, in a era overwhelmed by images and hyper-individualities such the one we are actually going through. Ideas are dialectic, open . . . shapes, on the contrary, are authoritarian and closed.

The only chance of comparison is, therefore, in the world of ideas, beyond both expressionism and linguistic rhetoric. In the same way project experiments new perceptions to establish between a man and a work, in order to create a more suitable and stimulating space for the subject. Each of our design ideas is weighed in relation to the project of perception and of the "look" that it will yield once it materializes.

Each project becomes an ethical-existential process around the evolution of an idea. In this way the project can gather and interpret a certain "shape of time," not as a linguistic problem of "shape of things," but as interpretation of perception of the subject that forms it, observed in its possible transformations and evolutions. For this reason, we do not support the existence of beautiful or ugly architectures, we rather believe there can be architectures that either have "sense" or not. Among the peculiarities of design opportunities, we try to face each project as a process of sense: at a certain point, and without particular effort, well-thought-out and researched architecture assumes a certain shape; shape is consequence of sense and not vice versa. Continuity of thought without continuity of shape. We are particularly attracted by one thing: reaching an almost primordial and synthetic sense of the subject, capable of producing a comeback of perception: the result is the willingness to reduce the presence of the subject "to the minimum," always finding the smallest unit of resistance, producing a continuous oscillation between weight, transparency, and opacity and also using the same material such as glass, observed in its various perceptions and in its reverberations. In our projects and, above all, in the perception process that the materialized work generates, we try to experiment "to the limit" the project theme in order to erase, as Merleau-Ponty would say, that "distance between us and the world," to reduce the "distance between reality and its perception" more and more.

n!studio访谈

在艺术与建筑之间的博物馆设计
博物馆设计中容器和内涵的关系

Antonello Stella

您如何对建筑进行定义？当代建筑实践意味着什么？

在Augustine看来，阐释建筑就跟阐释时间一样：如果没有人问起我，我就了解它，但一旦有人问起我，我又不了解它了。它就在我们的面前，任何一种解释似乎都是显而易见的。但是，如果我们想要搞清其深层次的含义，每一种解释似乎又不充分。杰出的建筑作品也许就是其纯“可见”层面的成功表现与出其“敏感”层面之后所能够立即表达出来的意义，而这同其简单的外表是不一样的。

我们可以说，当今的建筑仍然停留在“表观形式”的规范之内，而这种形式在如哲学或艺术等其他领域已经被超越或确切的来说正处于危机之中。城市生活的新形式同正在加速发生着巨大转变的区域生活新形式相比较，我们可以发现其在实际建筑形式方面有一种普遍意义上的不足感。在这种意义上，当明确了开发的灵活性、转化、成长以及建筑的暂时性，并重新思考这些项目开发的重要元素的存续性和合法性时，我们不相信任何详细阐述在更加相互关联、蔚为奇观的建筑形式或语言上的答复。在建筑项目中需要明确铭记的新核心，将在大量的个性特征中扩展开来，这样个体的视角可能会变得同分配和转化建筑空间的真实过程同样重要。

因而，我们现在便面临问题的第二部分内容，现如今进行建筑作业时所涉及到的更多的是道德规范问题，而不是审美学问题。很不幸，现如今的建筑业遵循的是审美学的规范，这与研究新型“道德规范”领域问题的哲学和艺术趋势恰恰相仿。建筑学不仅需要研究“事物的外形”，最终还需要研究一种能够指导并促进事物在转化和演变过程中“事物自身之间的关系”的遗传密码，甚至是预测其可能的“消失”情况。

意大利式建筑存在吗？如何更好地谈论它？什么最能表现其特色？

我们并不能像讨论西班牙、荷兰、法国或美国的建筑那样来讨论意大利当代建筑的情况。在20世纪80年代到90年代所发生的建筑文化重大否定活动，在我们国家之外很少有成功的，其所追求的是现代的连续性并以技巧和技术演变为特征，使得同国外环境相比其重要性丧失了很多。

在其他国家通过坚实的建筑体创造出了其特征，无论好坏，都能够代表其确切的“实体”，而意大利，在其近乎痴迷地寻求其独特“语言特征”的过程中，在其周围建造了一个实实在在的自我参考型“笼子”。最近，一种隐藏在假定“极简抽象主义”非干预之下的特定非项目文化正在全国范围内蔓延或重新显露风头。这就意味着将非常危险地把过去十年中我们这一代很多建筑师们的全部努力重新划归为零。在过去几十年中，建筑项目在“专业模式”和“文化模式”方面所出现的区分，在专业活动和大学活动之间所出现的区分，更进一步地影响了意大利的建筑景观格局。恰恰相反的是，一种看起来似乎能够代表年轻一代意大利研究特征的领域，也就是我们认为我们所属的领域，愿意直接面对真实的建筑世界，同时也面对着相同额度的理论研究，这种程度的理论研究一直以来都是意大利建筑思维的特色。而这肯定是学校所教授的知识中所没有的，与在描述语言方面的痴迷愿望相差很远。

正如我们在第一项回答中已经提到的，这种态度同一种新型“规划道德标准”也有关系。

新颖的，稳固的，舒适的，分享的，可辨识的，令人兴奋的，漂亮的……今天当您在设计建筑时，您更注重哪些价值和目标呢，原因是什么？

我们认为，这种作为项目“意义”的构思比其形式更能经受起长时间的考验，而且在如我们所处的这个充满影像和超个人主义的时代，只有通过这种方式我们才能够进行共同研究，就像我们所采用的一样。构思是辩证、公开的。而外形则恰恰相仿，是专制且封闭的。

因此，我们唯一进行比较的机会就是在构思世界中进行，在表达和语言修辞之外。按照同样的方式，针对需要在人员和工作之间建立新感知进行了试验，来为其主题创造一种更适宜、更有促进作用的空间。我们把我们每一个设计构思都与感知项目以及实现此构思之后其所能够呈现的“外观”进行对比，对其进行了评价。

每一个项目都变成了一种围绕某一构思的演变而进行的道德存续过程。这样，项目就能够聚集并诠释一种特定的“时间形状”，不将其诠释成一种“事物形状”的语言问题，而是作为对其形成主题的感知的诠释，正如在其可能转化和演变过程中所观察到的那样。因此，我们不支持美丽或丑陋建筑物的说法，我们更相信有具有“意义”的建筑物，也有不具有“意义”的建筑物。在设计机遇的特性中，我们努力将每一个项目看作是一个意义形成的过程：在某一特定点，不用作出特殊的努力，经过透彻构思和研究的建筑就会形成一种特定的外形。外形是意义的结果，而不是相反的过程。构思持续而外形不能持续。

某一件事对我们特别有吸引力：就是造就能够复原感知的几乎是原始和合成的主题意义。其结果就是愿意将该主题的存在性降到“最低”，总是遇到最小的阻力，在重要性、透明性和不透明性之间产生一种持续的摆动，也采用如玻璃等同样的材料，就像在其各种感知及其反映中所看到的那样。在我们的项目中，更重要的是，在物质化工作所生成的感知的过程中，我们努力尝试做到项目主题的“极限”，以越来越多地去除如Merleau-Ponty所说的“我们同世界之间的距离”，去除“现实及其感知之间的距离”。

Biography

Susanna Ferrini

Susanna Ferrini's approach to architectural design and research is focused on the concepts of transformation and evolution of architecture, confronted through the relationship between technique and architectural form. Her Research Doctorate, completed in Rome between 1993 and 1995, dealt with the interpretation of the development of a new concept of prefabrication, between the modern and the contemporary, understood as a form of architectural prefiguration. Concepts of flexibility, 'growth' and the evolution of architectural forms over time are also found in the work of Jean Prouvé, 'discovered' in 1990 and studied during a lengthy period of research in Paris made possible by a study grant from the French Government. During this period the industrial approach to technique was seen as the means of escaping from the trappings of formal research, allowing for a rediscovery of the essence of designing and reflecting upon the concept of temporality in architecture. Time and Form became the base of her reflections, surpassing the borders between technology and architectural research, between an industrial approach and the evolutional nature of architectural form. Matter and building intuition became the driving force for reflecting on the relationships between perception and forms of appropriating and transforming structures used by the inhabitants of our current landscape-world. The key is thus that of seeing buildings as industrial prototypes, capable of evolving and flexible towards transformation. The concept of 'envelope' becomes a system for interpreting the contemporary condition. In this field of experimentation, the new methods of dwelling become central to the development of conceiving of architecture in terms of its transformation and temporality, as a motor of transformation in progress. Over the years, her research into dwelling has focused on the theme of the industrial house during the modern and contemporary era and the mobile house as a representation of the social and cultural nomadism of our times. These themes were confronted in the publication of her Doctorate thesis entitled The architectural prefiguration in the dialectic between perfection and temporality. She has also presented this work at international conferences and in published articles; she was the editor of issue 20 – 21 of the magazine PianoProgettoCittà, published by the IDEA-DART Departments of the Faculty of Architecture in Pescara, which focused on the theme CasaNewMotion, New Methods of Dwelling (2004). Susanna Ferrini has taught at the Faculties of Architecture in Ascoli Piceno and Roma Tre. Since 2001 she is a professor in Architectural Design at the G. D'Annunzio University of Chieti, Faculty of Architecture in Pescara.

苏珊娜·费里尼（Susanna Ferrini）的建筑设计与研究模式着重于转化和演变的概念以及建筑的时间属性，其所面对的是技术和建筑形式之间的关系。她于1993~1995年间在罗马所完成的博士学位的研究诠释了在现代和当代进行预先构思这种新概念的发展，这种预先构思可以理解为为了在完成性和暂时性之间达到平衡而进行的一种建筑预想。在吉恩·普洛弗的作品中也体现了这种适应性的概念，即建筑形式随时间的“成长”而演变，吉恩·普洛弗在1990“发现”了这种概念并在随后借助由法国政府资助的研究基金在巴黎对其进行了长时间的研究。在此阶段，通过工业手段进行技术研究被认为是逃避正式研究的虚饰手段，使人们重新发现了在建筑中设计和反映暂时性概念的精髓。时间和形式成为了她在考虑工业方法和建筑形式的演变性特性时的构思基础，超越了技术和建筑研究之间的界限。对具体事物和建筑直觉的感知，成为了我们对分配和转化当代世界居民所使用的建筑结构之间的形式关系进行思考的驱动力。其关键就在于将建筑物看作是工业模型，认为它们能够进行演变并适于进行转变。而“信封”的概念则成为了对当代状况进行诠释的系统。在此试验领域，新的居住方法成为了在转变和暂时性方面进行建筑构思发展的核心，就像一台已经开始运行的转变电机。多年以来，她在住宅方面的研究集中在了现代和当代工业房屋，以及代表了我们所在时代的社会和文化游走性的移动式房屋这一主题。她在其所发表的题为La prefigurazione architettonica nella dialettica tra perfezione e tempora lit博士论文中阐述了这些主题来迎接公众的质疑，同时她在国际会议上和其所发表文章中呈出了此项研究。她曾担任Pescara建筑学院IDEA-DART系所发表的PianoProgettoCitt杂志20-21期的编辑，该刊集中探讨了住宅新方法（Casa New Motion）这一主题（2004）。苏珊娜·费里尼曾在Ascoli Piceno和Roma Tre的建筑学院任教。从2001年开始，她在Pescara的GD' Annunzio University of Chieti Faculty of Architecture任建筑设计学教授。

Antonello Stella

Born in Legnano in 1961, Antonello Stella graduated in Rome in 1990. His architectural design thesis was focused on the new rail stations of the Northern Rome Line. He later completed his Perfection Studies in Architectural Theory in Rome in 1991 and a Doctorate Research Degree in Architectural Composition in 1995, with a thesis entitled Tempi Ferroviari Tempi Urbani, Spazi Ferroviari Spazi Urbani, an exploration of the ties between infrastructure, the city and its processes of metabolisation of rail lands. This generated an understanding of railway design as a 'primary' form of designing the 'void'. In 1998 he obtained a Research position with the Faculty of Architecture in Ferrara, where he currently teaches Architectural Design. His research work is now focused on the relationship between infrastructure and the landscape. For the University of Ferrara he coordinated a research group that was part of the national inter-faculty project entitled Paesaggi Fluvio Marittimi, studying the river landscapes of the Po' Valley, its infrastructures and transformations. He is a member of the design group of the same faculty, for whom he has completed the new Faculty of Economics inside the Palazzo Bevilaqua Constabili, located in the historical centre and designed a new Study Centre in Cento. For the same group he studied the architectural and landscape implications of the conversion of the electrical power plant in Porto Tolle from combustible oil to coal. Antonello Stella has participated in numerous conferences exploring the theme of the reuse of historical heritage for museums in Rome, Cagliari and Hanoi.

Together with Susanna Ferrini he curated a monographic issue of World Architecture magazine, published by the Tsingua University of Beijing focused on the theme of museum architecture in Europe. During his Doctorate Thesis, further investigating the themes of the relationship between thought processes and architectural design, his interest in architectural theory overlapped with an interest in philosophical theory as a founding condition of meaning in our practice. This oblique condition informs both his teaching activities in Ferrara and professional practice in Rome. Though constantly immersed in architecture, in all of its multifaceted aspects, he seeks its meaning elsewhere, convinced that a great deal of architectural debate, with the exception of a purely technical branch, an inherent condition of the practice, in reality remains locked within an aesthetic paradigm that has now been surpassed by the ethical necessities of the contemporary condition.

安东内洛·斯特拉(Antonello Stella)出生于1961年，1990年毕业于罗马。其建筑论文集中研究了北罗马线的新火车站。随后他于1991年在罗马完成了其建筑学理论进修研究并于1995年完成了建筑布局学博士学位研究，其论文题目为Tempi Ferroviari Tempi Urbani, Spazi Ferroviari Spazi Urbani，文中研究了基础设施、城市及其火车网络新陈代谢进程之间的联系。这使他获得了将铁路设计作为其“空旷地”设计的“主要”形式的认知。1998年他在Ferrara的建筑学院获得了一份研究职位，目前他仍在此学院教授建筑设计。他的研究工作目前集中在基础设施和地貌风景之间的关系方面。在Ferrara大学他负责协调管理一个命名为Paesaggi Fluvio Marittimi的国家级教职员工研究项目一部分的研究团队，负责研究Po' Valley区河流景观、基础设施和转变。他同时也是该教职员工设计组的一员，并为其完成了位于Palazzo Bevilaqua Constabili内历史中心处的新经济学院项目并为Cento设计了一个新的研究中心。他还为同一研究组研究了将Port-Tolle的一座发电厂从燃油发电转化为燃煤发电的建筑和地貌风景意义。安东内洛·斯特拉曾参加了多项以罗马、卡利亚里和河内的博物馆研究挽救历史遗迹方案为主题的会议。他同苏珊娜·费里尼一起组织编辑了一期《世界建筑》杂志，该杂志由北京的清华大学主办，其主题集中在了欧洲的博物馆建筑。在他博士研究期间，他进一步研究了思维过程和建筑设计之间的关系这一主题，他在建筑学理论方面的兴趣同其在哲学理论方面的兴趣相互交融，成为了为我们的实践活动赋予意义的基础条件。他在Ferrara的教学活动又体现在罗马的专业实践活动中。虽然他一直沉浸在建筑学研究中，但其在所有多层面领域均会从其他角度寻求事物的意义，他确信参与对建筑学的辩论，除了在纯技术领域之外，都是我们实践活动中的固有条件，而在现实情况下它们都被限定在一种审美学范例范围之内，而目前这又被当前条件的道德需求所取代。

Office

MANAGEMENT TEAM: Susanna Ferrini Antonello Stella | DESIGN COORDINATION: Susanna Ferrini Antonello Stella Simone Bove | DESIGN TEAM: Stefano Battaglia, Anna Hollstein, Paolo Rossi, Angelo Venturi | STAFF Fabrizio Romano, Lara Turchini, Francesca Viglione, Mariana Mazzetta , Eleonor Court, Stefano Bulian, Matteo Ascani, Mario Russo, Sofia Steffenoni | CONSULTANTS: 3TI, PRAS Consulting, BETOM Ingenierie, Beterem Ingegnierie SA, Beterem Rennes, INCOM Ingegnierie, BCEOM Ingegnierie, ing. Roberto Braga, ing. Claudio D'Amico, ing. Vincenzo Pennacchi, arch. Fabrizio De Cesaris, Piergiorgio Rosati M&P, PMG Group,Umberto Pessolano Scientific Museography, Federico Marazzi Historical Museography, Dario De Luca Educational Expert, Claudia Mornati, Massimo Montanari, Giovanni Bulian, David Serero, Cristiana Marcosano Dell'Erba and Marco Biuzzi, Francesco Eleuteri, Carolina De Camillis Riccardo Fibbi Lighting Design, Roberto Creton Multimedia, Bernard Bellion Jourdan Landscape Designer, Daniela Moderini and Laura Zampieri Landscape Designer, Sara Sansalieu Landscape Design, Enrico Bonaiuti Graphic Design, Alessandra Montania Graphic Design, Lavinia Cozza Graphic Design, Marco Galofaro Model | STUDIO STAFF (1990 TO 2008): Andrea Antico, Barbara Annunziata, Chiara Barale, Stefano Battaglia, Fabio Bonfà, Simone Bove, Claudia Camplone, Dante Cecili, Andrea Ciofi degli Atti, Anna Cornaro, Maria Angela Corsi, Paolo D'Addato, Massimo D'Ascenzio, Thierry Decostered, Francesca Delicato, Valerio De Divitiis, Alessandra Di Franco, Birgitt Dittmar, Barbara Falcone, Claudia Farinelli, Francesca Federici, Fabrizio Federico, Sara Ferrazzoli, Nanette Föhr, Vito Fortini, Franca Francescucci, Davide Gamba, Paolo Iotti, Bertrand Lascourreges, Melanie Ligier, Giulia Mazzarelli, Mauro Mazzarelli, Roberto Meschini, Gabriele Misso, Tiziana Monti, Andrea Pasquato, Marco Pavarani, Simone Piacenti, Paola Pozzorini, Vincent Rapin, Jutta Riehle, Lara Rodriguez Perez Porro, Loris Rossi, Paolo Rossi, Giulia Scaglietta, Rocco Smaldone, Paolo Soellner, Camilla Thermes, Alessandro Turina, Angelo Venturi, Sole Zamponi, Teresa Zanaria, Piero Zinelli | PHOTOS: Alberto Muciaccia, Roberto Stevenson, Luigi Filetici, Alessandro Ciampi, n!studio.

Office Ex-Lanificio Luciani | Via di Pietralata 157-159 | 00138 | Rome Italy | T. +39677071723 | F. +39645436937 | email. nstudio@tiscalinet.it
www.nstudio.it

Credits

completed projects

1991-1993	PITIGLIANO (GR) Archaeological Museum inside the Orsini Fortress
1996-2000	NAZZANO (RM) Tevere River Museum inside the Savelli Fortress (XII-XV Sec)
1996-1999	NAZZANO (RM) Tevere River Museum Temporary Exhibition Centre
1996	VENICE n!studio Exhibition at VI Interantional Biennale of Architecture
1996-2000	ROME Public Spaces at the Olympic Village
1997	ROME North-Rome Railway Renovations
1997-2003	CIAMPINO (RM) Cultural Centre
1999-2007	CASTEL SAN VINCENZO (IS) Archaeological Museum
1999-2006	FERRARA New Faculty of Economics inside the Bevilacqua Costabili Palace
2000	CENTO (FE) University Cultural Centre, Feasibility Study
2001	PORTO-TOLLE (RO) Environmental Study for Enel Electric Power Station
2001-2002	ROME Archaeological Deposit and Scientific Laboratory inside the Archaeological Park of Villa dei Quintili
2001	CASTROVILLARI (CS) Project for a Public Space
2002	NAZZANO (RM) Ecomuseum inside Casale Bussolini
2002	CIVITAVECCHIA (RM) Enviromental Study for Enel Electric Power Station Torre Valdaliga
2002	CATANZARO Public Urban Park along the Fiumarella River
2002	CODIGORO (FE) Suspended Water Tank
2002-2004	NAZZANO (RM) Fossils Museum Section
2003-2007	FORLIMPOPOLI (FO) Cultural Centre Casa Artusi "La Biblioteca in Cucina"
2004	PITIGLIANO (GR) Exhibition Centre inside Orsini Fortress
2004-2006	ROMA New Library inside the Ss.Quattro Coronati Convent
2004-2006	PITIGLIANO (GR) Ragnini Private House
2005	NAZZANO (RM) Exhibition Centre and Tourist Resort inside the Savelli Fortress (XII-XV Sec)
2005-2006	BEIJING (CHINA) Exhibition "Mirroring The Age", Millenium World Art Museum
2006-2006	BEIJING (CHINA) Exhibition "The Great Civilizations" Millenium World Art Museum
2006-	ROME Nievo and Verbano Nemorense Subway Station for the new D Line Preliminary Project
2006-	BRETEIL (FRANCE) Archaeological Museum and Research Centre De La Breche Et De La Noye
2007	ROME Ugo Mulas Exhibition at Maxxi Museum
2008	ROME Exhibition "Pensare la Materia-n!studio projects" at Come Se Gallery

ongoing projects

2005	ROME New Office Building in Via Pallacorda
2006	NAZZANO (RM) Parking Lot and Public Services
2006	BEIJING (CHINA) Exhibition for RM008
2006	BEIJING (CHINA) Proposal for a New Atrium of the Capital Museum, Beijing
2007	ACILIA (RM) Restoration of Rural Houses in Giano Centre
2007	ISOLA DEL GIGLIO (GR) Civic Museum
2007	VITERBO Thermal Centre and Resort
2007	ROME Ecoromabuilding Residential Tower
2007	ROME-AXA Renovation of the ex-Drive in Area
2008	XI'AN (CHINA) Western Market Museum
2008	PADOVA Mixed-use Tower with Residences Offices Facilities
2008	GAP (FRANCE) Solenha Solareuromed Thermosolar Power Station Headquarters

competitions

1992	LODIVECCHIO (MI) Public Housing, mention
1992	AGUZZANO (RM) Public Space of Access to Urban regional Park, 3rd price
1993	MARCON (VE) Public Housing, mention
1995	ROME Borghetto Flaminio, 6th price
1996	ROME Centopiazze Neighbourhood Piazzas, mention
1997	BERGAMO Housing
1998	VITERBO Restoration of the ex-Slaughterhouse 2th price
1999	MILAN "Five Piazzas for Milan"
2000	TURIN Porta Susa Raylway Station
2005	VIENNA (AUSTRIA) Addiction to Kunsthistorische Museum
2007	ITALY Small Railway Stations
2007	SALAMANCA (SPAIN) Museum of Architecture and Urban Planning inside "Banco De Espana" building
2007	BEIJING (CHINA) GBD Art District 500m3 Housing Units

shortlisted competitions

1992	PENNE (PE) Contemporary Art Museum, winner
1997	ROME Renovation and Addiction of the National Modern Art Gallery GNAM
1999	SALISBURGO (AUSTRIA) Renovation and Addiction of Carolino Augusteo Museum
2000	BRESCIA Centre for Contemporary Culture at the "Crocera di S. Luca"
2000	BRESCIA Tarello Park
2001	MURANO (VE) Restoration of the "Conterie" building in Venice
2001	ROME Renovation of Piazza Augusto Imperatore
2002	AIX EN PROVENCE (FRANCE) Archaeological Deposit and Laboratory
2004	ILE DE LA RÉUNION (FRANCE) Centre for Water Treatement of a Salt Lake in Saint-Pierre
2004	MARSEILLE (FRANCE) Conservation Centre for the Museum of European Civilizations of the Mediterranean
2004	ROME Three New School in Casal Monastero
2004	AMALFI (NAPLES) "Valle Dei Mulini" Tourist Cultural Centre
2004	ITALY Ance"Housing in the Future"
2005	NAPLES Three Restaurants for Naples Museums
2005	TOULON (FRANCE) Museum of Natural History
2005	ALBY SUR CHERAN (FRANCE) Museum of the River winner ex-aequo
2005	CAMPOBASSO Addiction to the "A.cardarelli" Hospital
2005	MONTAUBAN (FRANCE) Conservation Centre for the Museum Ingres
2006	BRETEIL (FRANCE) Archaeological Museum and Research Centre De La Breche Et De La Noye winner
2006	RENNES (Francia) Ecomuseum in Rennes
2006	ROME Kindergarden Pre-school and Public Park
2006	GAP-ASPRES SUR BUECH (FRANCE) Maison De La Lumiere
2007	BARI Provincial Archaeological Museum
2008	BREMERHAVEN (GERMANY) Addiction to the Maritime Museum
2008	BERLIN (GERMANY) Addiction to the City Museum

publications

1991 Napoli Architettura e Città
"L'Area della Floridiana"

1992 Attraversare la Città dal Porto al Passetto
"Un'Idea per Ancona"

1992 Archeologia Viva
"Pitigliano,Come Nasce un Museo"

1992 Exhibition "Arte e Architettura" Catalogue
n!studio, Progects for Penne

1993 Public Space in Small Cities
"Nove Progetti per Penne"

1993 Fuori Uso-Architettura Italiana-Un Confronto fra Generazioni
Archaeological Museum in Pitigliano

1993 D'Architettura N°10
Giangiacomo D'Ardia, "n!studio"

1993 Edilizia Popolare" N°226
Massimo Bilò, Comments to the Competition "Opera Prima",Lodivecchio (Mi)

1993 Segno N°121
Oscar Buonamano, Progects for Penne
Contemporary Art Museum in Penne

1994 Progetto Ortona-La Manutenzione del Territorio:the Port Area Design
Susanna Ferrini "Artificiale, Costruito, Ricostruibile"
Antonello Stella "L'Edificio Inesistente"

1994 Do.Co.Mo.Mo-Conference Proceedings
Susanna Ferrini "Industrial Thought on Technique as Seen Through the Works of Jean Prouvè"

1994 Thoughts and Form, Time and Design

1994 Edilizia Popolare N°233
Franco Zagari, Competition for the Urban Regional Park of Aguzzano

1994 Edilizia Popolare N°234
n!studio-Design Text, Marcon (Ve) "Opera Prima"

1994 Industria delle Costruzioni N°277

1994 Exhibition "Fuori Uso" Catalogue
Pino Scaglione, Contemporary Architecture: Comparison among Generations

1995 New S. Filippo Neri Church, Notes for a Contemporary Project

1995 Segno N°142
Rita Olivieri, "Osservatorio di Confine"

1995 Industria delle Costruzioni N°287
Domizia Mandolesi, Competition for the Urban Regional Park of Aguzzano

1995 Exhibition "Pensiero e Forma" Catalogue
n!studio-Projects

1996 D'Architettura N°16
Archaeological Museum In Pitigliano and Museum Complex in Nazzano

1996 Catalogue of the VI Venice Biennale of Architecture
Exhibition "Sensori del Futuro/L'architetto Come Sismografo" at the Italian Pavillon

1996 Segno N°151
Oscar Buonamano, VI Venice Biennale of Architecture
The Italian Partecipation

1996 Casabella N°639
Projects for Rome

1997 Hundred Idea for Piazza Oberdan Competition,
Catalogue of the Exhibition

1997 Controspazio N° 3/97
Paola Veronica Dell'Aira, Projects for Rome
Public Spaces at Olympic Village

1998 Casabella N° 656
Pippo Ciorra, Architecture-Italy

1998 Ricerca e Progetto N.11-12
Competition Borghetto Flaminio in Rome

2000 Il Mondo N. 5
Franco Stefanoni, Avanti Piano ma non Troppo

2000 Il Progetto N°6
Franco Zagari, Gerico 20+20+20

2000 Giuseppe N°8
n!studio, Cronaca: my Profession

2000 Corriere Della Sera 1/11
A New River Museum in Nazzano

2001 Industria della Costruzioni N°356/2001
Ippolito Pizzetti, River Museum in Nazzano

2001 Darc (General Direction for Contemporary Art and Architecture)
D40 - Ideas for a new Info Space
n!studio, Diaframmi di Luce Solida

2001 Costruire N°215
Antonella Marino, Occasione Borromini

2001 Architettura e/è Arte
n!studio, La Forza del Pensiero

2001 Metamorfosi-Quaderni di Architettura e Arte N°42-43 Alessandra Muntoni, 12 Giovani Gruppi Romani, la Ricerca della Trasversalità

2001 Spazio d'Autore N°16 a Project from A to Z
Nazzano

2001 Rivista del Vetro N°8
M.g. Marchelli, Nazzano River Museum-Simulated Tevere

2002 D'Architettura N°18-Cento Progetti-Observatory on Italian Construction
Archaeological Museum of San Vincenzo al Volturno

2002 50/50 Nuova Architettura Italiana
Two Generations Compared, n!studio, Archaeological Museum in Pitigliano and Scientific Centre and Museum in Nazzano

2002 La Guida dell'Ufficio Tecnico N°1
Luigi Prestinenza Puglisi
"Lungo il Fiume è Nato un Museo"

2002 Il Progetto N°11 Luca Molinari, Italia 2001 Austellung Italienische Architektur

2002 Costruire N°228

Antonella Marino, Relazioni Percettive

2002 Spazio Architettura N°55
n!studio, Nazzano River Museum, Pitigliano Archaeological Museum, Garden in Rome

2002 Domus N°849
Luigi Prestinenza,Puglisi, Racconti del Fiume

2002 Dal Futurismo al Futuro Possibile nell'architettura Italiana Contemporanea
n!studio, Nazzan River Museum

2003 Bimestrale dell'Ordine degli Architetti di Roma e Provincia-Special Issue New Exhibition Spaces in Rome, Nazzano River Museum

2003 Catalogue "Gold Metal for Italian Architecture"
n!studio, Nazzano River Museum

2003 Architecture Interieure Cree - Avril-Mai N°308"
Ecosystème Muséal

2003 Arte/Architettura/Città
38 Progects for Piazza Augusto Imperatore in Rome
n!studio, for a Open Air Museum

2003 L'Espresso N°Novembre pag.16
"San Vincenzo non fa Miracoli"

2004 Progettare N°18
Valerio Paolo Mosco "La Casa del Futuro"

2004 Industria delle Costruzioni N°375
Valerio Mosco, Archaeological Museum of San Vincenzo al Volturno

2004 Ottagono N°169
Luigi Prestinenza Puglisi, Leggerezza Vetrosa per n!studio

2004 Ar2 N°2
Susanna Ferrini e Antonello Stella, Passando per Konigsberg

2004 Vanity Fair N°36
Luca Molinari-Il Container Futuristico d'Acciaio Arrugginito, Storage for Archaeological Relics inside "Villa dei Quintili" in Rome

2004 Sumau N°11-12
Miwako Nagamachi, River Museum

2005 Architetti-Idee Cultura e Progetto
Water Power: la Valle dei Mulini ad Amalfi

2005 Tre Punti di Ristoro per Tre Musei Napoletani
Restaurant Area inside Capodimonte Museum

2005 Verso una Nuova Qualità dell'Abitare Ance
Competition for the "House of Future"

2005 Il Giornale dell'Architettura N°25
Michele Bonino-Un'Architettura Necessaria

2005 Progettare-Bauwelt N°1-2
Bauwelt Preize 2005 Storage-Laboratory for Archaeological Relics inside "Villa dei Quintili" in Rome

2005 Industria delle Costruzioni N°380
Domizia Mandolesi, Architettura Forma Materiali
Storage-Laboratory for Archaeological Relics inside "Villa dei Quintili" in Rome

2005 Industria delle Costruzioni N.381
Valerio Paolo Mosco, Nonostante una Tendenza Generalizzata,Conservation Centre for the Mediterranean Museum in Marseille

2005 Architetti Italiani Under 50 – Ricerca Formazione Progetto di Architettura
Adelaide Di Michele, Nazzano River Museum,

2005 Net.it-a Snapshot of Contemporary Architecture, Design and Photography
"n!studio-Progetti". Conservation Centre for the Mediterranean Museum in Marseille

2005 Giornale dell'Architettura N°25, Janvier
Giorgio Santilli-La Nuova Architettura

2005 Giornale dell'Architettura N°33 Octobre
Luca Gibello-Holcim Award to an Italian Project
The Water Mills Valley in Amalfi

2005 Wa-World Architecture N°184 Octobre 2005
Nazzano River Museum, Archaeological Museum San Vincenzo al Volturno

2006 Spazi Pubblici Contemporanei, Architettura a Volume Zero, Suspended Water Tank in Codigoro (Fe)

2006 D'Architettura N°29 Gennaio/Aprile 2006
Archaeological Museum San Vincenzo al Volturno

2006 Domus Special Issue N°897 November 2006
"Architettura di Acqua e di Luce", Suspended Water Tank

2006 Paesaggio Urbano N°2
Waterpower:The Water Mills Valley in Amalfi

2006 Giornale dell'Architettura", N°38 Marzo 2006
Exhibition "Mirroring the Age" Pechino

2006 Wa-World Architecture N°195 – 09/2006
Deposito-Laboratorio di Reperti Archeologici alla Villa Dei Quintili. Mostra "Mirroring the Age" Beijing

2006 Gold Metal for Italian Architecture
Storage-Laboratory for Archaeological Relics inside "Villa dei Quintili" in Rome

2006 Il Sole24 Ore Edilizia e Territorio-Progetti Concorsi N°22 June 2006 Archaeological Museum in Breteuil

2006 Arie Italiane-Motivi dell'Architettura Recente Edizioni Edilstampa 2006
Archaeological Museum San Vincenzo al Volturno

2006 Progetti Roma -n°3, 2006
Cultural Centre in Ciampino

2006 Il Sole24 Ore Edilizia e Territorio-Progetti Concorsi N°16 October 2006
Suspended Water Tank in Codigoro (Fe)

2006 Tc Quadernos N°74
Archaeological Museum in Breteuil

2006 Metamorfosi-Quaderni di Architettura" N° 62,
Exhibition "Mirroring the Age" Beijing

2006 Paesaggio Urbano N°2 2006
Luigi Centola, Exhibition Centre and Mediterranean Products Shop inside the ex Soap Factory in Amalfi

2006 Decorate Magazine
Storage-Laboratory for Archaeological Relics inside "Villa dei Quintili" in Rome

2006 Architectural Review
Corten Gem Storage-Laboratory for Archaeological Relics inside "Villa dei Quintili" in Rome

2006 Staal.acier Edition Speciale 2006
Storage-Laboratory for Archaeological Relics inside "Villa dei Quintili" in Rome

2006 Roma, La Nuova Architettura
Storage-Laboratory for Archaeological Relics inside "Villa dei Quintili" in Rome

2006 Contemporary Ecologies. Energies for Italian Architecture
Waterpower: The Water Mills Valley in Amalfi

2007 Il Sole24 Ore Edilizia e Territorio N°25, 11 June 2007
Casa Artusi

2007 Il Sole24 Ore Edilizia e Territorio N°26, 2 July 2007
Exhibition Centre in The Orsini Fortress in Pitigliano

2007 Italy Now? Country Positions in Architecture curated by Alberto Alessi
Archaeological Museum San Vincenzo al Volturno

2007 Artusi. Quaderno N.1
Casa Artusi
2007 Detail N° 7/8 2007
Storage-Laboratory for Archaeological Relics inside "Villa dei Quintili" in Rome
2007 Il Giornale dell'Architettura N°53
Casa Artusi
2007 Architektur Aktuell
Storage-Laboratory for Archaeological Relics inside "Villa dei Quintili" in Rome
2007 Architetture, Città e Territorio
Contemporary Architecture in Tuscan Landscape
"Necessità del Contemporaneo,Interventi a Pitigliano"
2008 n!studio - Pensare la Materia
Monograph book about n!studio
2008 Domus N°913 Food April 2008
"A Casa con l'Artusi" project Casa Artusi
2008 Sustainab.Italy-An Overview of Contemporary Italian Archtecture Focusing on Sustainability
Ecoromabuilding Residential Tower in Rome
2008 Progetti N°7 Roma
Ecoromabuilding Residential Tower in Rome
2008 XI Venice Biennale of Architecture Exhibition Catalogue
2008 Catalogue of the Exhibition "Uneternal City.Thirty years from Roma Interrotta" at the XI Venice biennale of Architecture

exhibitions

1993 Projects for Penne, Central Library of Pescara School of Architecture
1995 Osservatorio di Confine
Contemporary Artists and Architects Compared
Città Della Pieve (Pg), Chiusi (Ar), L'aquila
1995 Pensiero e Forma Contemporary Artists and Architects Compared
Rome, St. Orsola Istitute
1996 Lungo il Fiume
Rome Province, Rome Municipality, Lungotevere della Vittoria
1996 La Fabbrica di Roma
Comune di Roma, Palazzo delle Esposizioni
Exhibition curated by Ufficio Cento Piazze of Rome Municipality
1996 VI Venice Biennale of Architecture
Giardini della Biennale, Italian Pavillon
1996 La Fabbrica di Roma
Piani e Progetti per il Rinnovamento Urbano Sezione Informativacasaidea '97 (1-9 Marzo), Fiera di Roma
1997 "Partito Preso - Architettura"
"L'Architetto e l'Artista a Confronto su un Tema Emblematico:l'Ampliamento di Spazi e di Funzioni della Galleria Nazionale d'Arte Moderna"
National Gallery of Modern Art, Rome
1997 Triennale of Milan
Palazzo of Triennale-Milan
1999 International Biennale of Young Artists form Europe and the Mediterranean
Rome, Testaccio ex Slaughterhouse
Exhibition Curated by Rome Municipality, Inarch
2000 Gerico. Sessanta (20 + 20 + 20) Young Italian Architects.
Prague, San Bartolomeo Chapel
Exhibition Curated by Italian Institute of Culture
2000 n!studio – Spazio d'Autore
Facolty of Architecture Rome
2002 50 New Italaian Architecture
Two Generations Compared
Haus Der Architektur, Graz, Austria
Exhibition Curated by Aid'a
2002 Italian Architects - International Exhibition of Art and Architecture Tokio.
Exhibition Curated by Ordine Nazionale degli Architetti-Ministry of Foreign Affairs and Culture
2002 Italian Architects - International Exhibition of Art and Architecture Bruxelles.
Exhibition Curated by Ordine Nazionale degli Architetti-Ministry of Foreign Affairs and Culture
2003 Milan Triennale-Gold Medal for Italian Architecture
Palazzo of Triennale-Milan
2004 Mon.it.orp-Works by Roman Architects
Casa dell'architettura-Rome
Exhibition Curated by Ordine degli Architetti di Roma e Provincia
2005 Gold Medal for Italian Architecture
The Urban Development Centre-Singapore
Triennale of Milan-Italian Chamber of Commerci
2005 Italy Now
Cornell University-Ithaca-New York, Curated by Alberto Alessi
2005 Architetti Italiani Under 50
Palazzo of Triennale-Milan
2006 Projects for Rome
Fondazione Olivetti-Rome
2006 Mon.it.orp-Works by Roman Architects
Acquario Romano
2007 "Recuperare il Passato:Sfide e Opportunità nella Conservazione e Valorizzazione del Patrimonio Culturale"
Italian Embassy in Hanoi
2008 Sustainab.Italy- London
2008 XI International Venice Biennale of Architecture
"Out There:Architecture Beyond Building"
2008 "Pensare la Materia-n!studio projects" at Come Se Gallery, Rome

lectures

1994 Projects for the Urban Regional Park of Aguzzano
Ordine degli Architetti di Roma
1998 Young Architects Compared
Faculty of Architecture of Ascoli Piceno
1999 Ar.id. Ideas in Built Reality
Losanna (Svizzera), Ar. Id. Curator
2000 Inarch – Italian Architects at the Biennale of Venice
Italaian Pavillon
Inarch (Ist. Naz. Di Architettura)
2003 Cornell University in Rome n!studio Recent Projects
Cornell University a Roma
2004 Faculty of Architecture "Ludovico Quaroni" Rome
n!studio Recent Projects
2005 Faculty of Architecture of Naples "Federico II"
n!studio Recent Projects
2005 Faculty of Architecture of Roma Tre.
n!studio Recent Projects
2006 Architecture in Rome Today
Casa dell'architettura, Rome
2006 Inarch - I Lunedì dell'architettura. "Pensare la Materia"
Associazione Nazionale Costruttori Edili a Roma
2006 Pensare la Materia-n!studio Projects
Ordine degli Architetti Pianificatori e Paesaggisti di Cesena
2006 Pietra e Creatività
Rapolano Terme (Siena)
2006 Great Museums in small cities
2006 Architettura. Fare Trasmettere Comunicare
Cesena
2006 La Città nella Città
Palazzo dei Congressi-Roma
2007 Recuperare il Passato:Sfide e Opportunità nella Conservazione e Valorizzazione del Patrimoio Culturale
Hanoi-Vietnam
2007 Vivere L'ambiente:Città e Territorio tra Crescita e Sostenibilità
Hanoi-Vietnam
2008 Pensare ad un Futuro Cromatico in Architettura
Bari
2007 1°Convegno Internazionale di Studi

	Architettura Contemporanea nel paesaggio Toscano Colle Massari-Cinigiano (Gr)
2008	Pesentation of the book "n!studio Pensare la Materia" Università La Sapienza Roma
2008	Workshop "Il Progetto dell'Allestimento Museale. Il Proceso Ideativo e la sua Realizzaizone" Galleria "Come Se"-Roma

awards

1993	Exhibition-Award "Fuori Uso '93" Ex Gaslini Buildings - Pescara "Under 40 - Occasioni di Sperimentazione" Pitigliano Archaeological Museum, winner
2000	Inarch Award "Opera Prima" Castel S.Angelo Rome, Inarch-Lazio Pitigliano Archaeological Museum, mention
2001	Inarch Award "Maurizio Sacripanti" Castel S. Angelo Roma Inarch-Ordine Architetti di Roma Nazzano River Museum, winner
2003	Gold Metal for Italian Architecture-Triennale of Milan-Ordine Nazionale degli Architetti Nazzano River Museum, mention
2004	Monitor/p Storage-Laboratory for Archaeological Relics inside "Villa dei Quintili" in Rome, mention
2005	European Holcim Awards Holcim Forum for Sustainable Construction Tourist Centre in the Water Mills Valley in Amalfi, winner
2005	World Holcim Awards Holcim Forum for Sustainable Construction Tourist Centre in the Water Mills Valley in Amalfi, 2nd price
2005	Bauwelt Prize Bauwelt Magazine Storage-Laboratory for Archaeological Relics inside "Villa dei Quintili" in Rome, mention
2006	Monitor/p Cultural Centre in Ciampino, mention
2006	Gold Metal for Italian Architecture-Triennale of Milan-Ordine Nazionale degli Architetti Storage-Laboratory for Archaeological Relics inside "Villa dei Quintili" in Rome, mention
2006	Architecturale Review Awards For Emerging Architecture Honourable Mention - Storage-Laboratory for Archaeological Relics inside "Villa dei Quintili" in
2008	Amate L'architettura Award Nazzano River Museum, winner

photo credits:

Alessandro Ciampi pag. 14, 15, 16, 17, 18, 19, 21, 176, 177, 240, 241, 242, 243, 244, 245, 246, 247, 249, 251, 253, 254, 255, 257, 258, 259, 260, 261, 263, 264, 265, 266, 267, 269, 270, 271, 272, 273, 274, 275, 276, 277.
Luigi Filatici pag. 99, 105, 109, 172, 173, 174, 175, 178, 179, 180, 181
Roberto Stevenson pag. 182, 184, 185
n!studio pag. 4, 44, 45, 46, 47, 48, 49, 50, 51, 52, 53, 54, 55, 212, 213, 301.
Antonello Stella pag. 294, 295, 296, 297.
Roberto Galasso pag. 186, 187, 193, 194, 195.
pag. 278-279 Jean Prouvé, Maison Coque, ADMM
pag. 280 Jean Prouvé, Maison Tropicale, ADMM
pag. 281 Jean Prouvé, Maison de Vacance, ADMM
pag. 282 Gerrit Rietveld
pag. 283 Casabella, Elemond s.p.a.
pag. 306 Emiliano Larizza

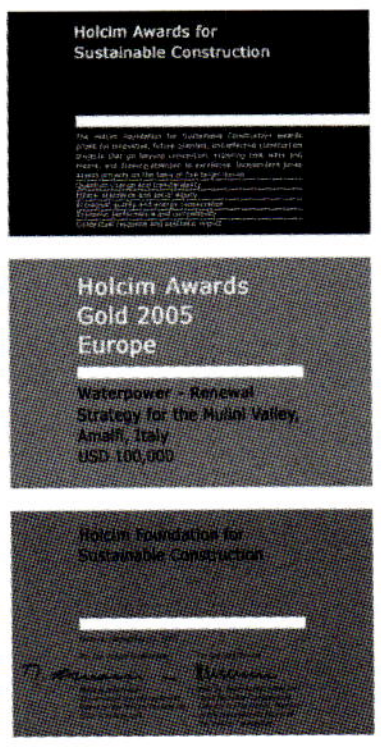

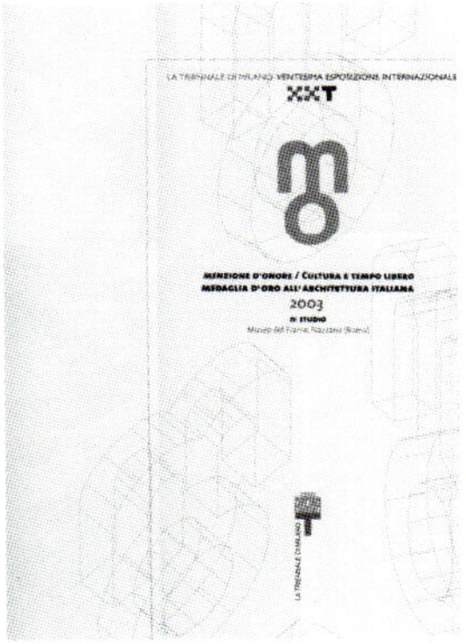

图书在版编目(CIP)数据

物质感官·生态罗马：n!studio作品 / 蓝青主编; 美国亚洲艺术与设计协作联盟
–武汉：华中科技大学出版社，2008.8
ISBN 978-7-5609-4835-5
I.物... II.①蓝... ②美... III.建筑设计—作品集—意大利—现代 IV.TU206
中国版本图书馆CIP数据核字(2008)第128417号

物质感官·生态罗马：n!studio作品
主　　编：蓝青
美国亚洲艺术与设计协作联盟(AADCU)

出版发行：华中科技大学出版社
地　　址：武汉市珞喻路1037号(邮编：430074)
出 版 人：阮海洪
责任编辑：张若琳
责任监印：张正林
制版印刷：北京画中画印刷有限公司
开　　本：889mmx1194mm　1/16
印　　张：19.50
字　　数：221千字
版　　次：2008年8月第1版
印　　次：2008年8月第1次印刷
ISBN 978-7-5609-4835-5/TU·402
定　　价：268.00元

销售电话：022-60266190，60266199(兼传真)
网　　址：www.hustpas.com